# Data Integration
*The Relational Logic Approach*

# Synthesis Lectures on Artificial Intelligence and Machine Learning

**Editors**
**Ronald J. Brachman,** *Yahoo! Research*
**Thomas Dietterich,** *Oregon State University*

**Data Integration: The Relational Logic Approach**
Michael Genesereth
2010

**Markov Logic: An Interface Layer for Artificial Intelligence**
Pedro Domingos and Daniel Lowd
2009

**Introduction to Semi-Supervised Learning**
Xiaojin Zhu and Andrew B. Goldberg
2009

**Action Programming Languages**
Michael Thielscher
2008

**Representation Discovery using Harmonic Analysis**
Sridhar Mahadevan
2008

**Essentials of Game Theory: A Concise Multidisciplinary Introduction**
Kevin Leyton-Brown, Yoav Shoham
2008

**A Concise Introduction to Multiagent Systems and Distributed Artificial Intelligence**
Nikos Vlassis
2007

**Intelligent Autonomous Robotics: A Robot Soccer Case Study**
Peter Stone
2007

Data Integration: The Relational Logic Approach
Michael Genesereth


ISBN: 978-3-031-00422-3  paperback

ISBN: 978-3-031-01550-2  ebook


DOI: 10.1007/978-3-031-01550-2

A Publication in the Springer series

*SYNTHESIS LECTURES ON ARTIFICIAL INTELLIGENCE AND MACHINE LEARNING*

Lecture #8

Series Editors: Ronald Brachman, Yahoo! Research, and Thomas Dietterich, Oregon State University

**Series ISSN**

ISSN 1939-4608        print
ISSN 1939-4616        electronic

# Data Integration
*The Relational Logic Approach*

**Michael Genesereth**
Stanford University

*SYNTHESIS LECTURES ON ARTIFICIAL INTELLIGENCE AND MACHINE LEARNING #8*

# ABSTRACT

Data integration is a critical problem in our increasingly interconnected but inevitably heterogeneous world. There are numerous data sources available in organizational databases and on public information systems like the World Wide Web. Not surprisingly, the sources often use different vocabularies and different data structures, being created, as they are, by different people, at different times, for different purposes.

The goal of data integration is to provide programmatic and human users with integrated access to multiple, heterogeneous data sources, giving each user the illusion of a single, homogeneous database designed for his or her specific need. The good news is that, in many cases, the data integration process can be automated.

This book is an introduction to the problem of data integration and a rigorous account of one of the leading approaches to solving this problem, viz., the relational logic approach. Relational logic provides a theoretical framework for discussing data integration. Moreover, in many important cases, it provides algorithms for solving the problem in a computationally practical way. In many respects, relational logic does for data integration what relational algebra did for database theory several decades ago. A companion web site provides interactive demonstrations of the algorithms.

# KEYWORDS

data integration, query folding, query optimization, ontologies, reformulation, computational logic

# Contents

# Preface

Data integration is a critical problem in our increasingly interconnected but inevitably heterogeneous world. There are numerous data sources available in organizational databases and on public information systems like the World Wide Web. Not surprisingly, the sources often use different vocabularies and different data structures, being created, as they are, by different people, at different times, for different purposes.

The goal of data integration is to provide programmatic and human users with integrated access to multiple, heterogeneous data sources, giving each user the illusion of a single, homogeneous database designed for his or her specific need. The good news is that, in many cases, the data integration process can be automated.

This book is an introduction to the problem of data integration and a rigorous account of one of the leading approaches to solving this problem, viz., the relational logic approach. Relational logic provides a theoretical framework for discussing data integration. Moreover, in many important cases, it provides algorithms for solving the problem in a computationally practical way. In many respects, relational logic does for data integration what relational algebra did for database theory several decades ago.

Chapter 1 introduces the problem of data integration and the relational logic approach to solving it. Chapter 2 provides mathematical background necessary for the main body of the book. Chapter 3 defines the key subproblem of query folding, describes one particular method in detail, and analyzes the problem of query folding in general. Chapter 4 describes the various issues involved in query planning, i.e., optimization, source selection, and execution planning. Chapter 5 discusses the issues involved in master schema formulation and reformulation.

One of the features of this volume is the availability of a reference implementation for all of the algorithms discussed in the book. The implementation is used directly in the demonstrations and exercises available in the interactive edition, and the code is printed in lightly documented form in the Appendix of both editions. Information on the interactive edition may be found on page xi.

This book is a compilation of material from a variety of papers on data integration published over the years. The compilation was refined by several offerings of a Stanford University course on data integration taught at the advanced undergraduate and early graduate student level.

In creating the book, I benefited from the generosity of many folks. Oliver Duschka, my co-author on various papers on data integration, kindly consented to allow me to publish our joint work here, in many cases, verbatim. Rada Chirkova, another collaborator, helped by talking with me about the material and sharing some of her latest work on database reformulation. Finally, Mike Morgan, the publisher, played a key role by repeatedly pestering me about progress in what, at the time, seemed to be indecently frequent intervals but which, in retrospect, demonstrated exemplary patience. Without his flexibility and encouragement, the book would never have come into existence.

# Interactive Edition

An interactive edition of this book can be found at:

http://logic.stanford.edu/dataintegration/

The interactive version makes available a reference implementation for all of the algorithms discussed in the book. The implementation is used directly in the demonstrations and exercises available in the interactive edition.

# CHAPTER 1

# Introduction

## 1.1 DATA INTEGRATION

Today's computer networks provide us with access to a wide variety of databases. There are company directories, product catalogs, airline schedules, government databases, scientific databases, weather reports, patient records, drug studies, and many other sources of "structured" data.

In some cases, the answers to our questions are contained in a single database; in such cases, it is easy to get those answers—a single request per question is all that is needed. More often, however, the necessary data is distributed over multiple sources. In such cases, *data integration* is necessary to combine data from these sources.

The good news is that, in many cases, the data integration process can be automated. This is the job of *data brokers*, i.e., online services that automate the data integration process. The goal of a data broker is to provide its clients with integrated access to multiple, heterogeneous data sources, giving each client the illusion of a single, homogeneous database designed for his specific need.

The architecture of a typical broker-based data integration system is shown in Figure 1.1. The boxes at the top of the diagram represent clients. Those at the bottom represent sources. The box in the middle is the data broker. In operation, the clients send queries to the broker; the broker interrogates the sources; the sources reply; and the broker sends the integrated results back to the clients.

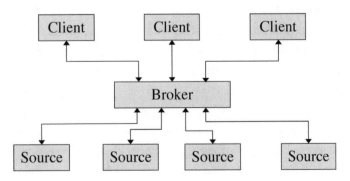

FIGURE 1.1: Data broker.

*An interactive version of this book is available that includes implementations of algorithms used in demonstrations and exercises. See page xi for more information.*

In this case, there are just four sources and three clients. In practice, the numbers are likely to be much larger. There can be thousands of sources and millions of clients.

Note that in the architecture shown, the clients do not interact with sources directly. They communicate only with the broker, and the broker communicates with the sources. In this way, data brokers differ from search services like Google. While Google returns a list of document references that the user must then examine for himself, a data broker extracts all relevant information and provides the user with a single answer to his query.

The online character of data integration through data brokers differentiates it from offline approaches, such as ETL (extract, transform, and load). The ETL approach involves extracting data from sources, transforming it to fit operational needs, and ultimately loading it into the end target, e.g., a data warehouse. One downside of this approach is that the data accessed by the users can be out of date. Moreover, by forcing all data into a limited common schema, information can sometimes be lost. By contrast, data brokering offers integrated access to live data. Also, it preserves full information; when done correctly, no data is ever lost in the process.

## 1.2 HETEROGENEITY

While collecting, aggregating, and disseminating data is the primary business of a data broker, this task would be relatively easy if not for differences among different data consumers and different data sources. These differences come in two sorts—format heterogeneity and conceptual heterogeneity.

*Format heterogeneity* has to do with differences in communication protocol and data format. The clients can be human users interacting through web forms (encoded in HTML); they can be application programs treating the broker as a virtual database (e.g., using ODBC); and they can be data warehouses using the broker to update their data. The sources can be relational databases (accessed using SQL), files in various formats (tab-delimited text, XML, etc.), and application programs (using LDAP, ADAP, etc.).

For our purposes here, it suffices to treat all sources as relational databases and all clients as users of relational databases. The reason for this is that we can handle other types of clients and sources by interposing *wrappers* to translate their communication protocols and data formats into a communication protocol and format appropriate to relational databases. While this is burdensome, given that there are only a small number of cases to be handled, the problem of creating such wrappers is manageable.

The main difficulty in data integration (and the focus of our attention in this book) is *conceptual heterogeneity*, i.e., differences in the schemas and vocabularies used by different data consumers and different data sources. Examples of such differences abound, from simple cases of different words and incompatible units of measurements to more complex cases involving different relational attributes and different relational tables. In order to use heterogeneous sources to answer questions from heterogeneous consumers, brokers must not only expend the energy to find and communicate with the sources but also deal with conceptual heterogeneity among those consumers and sources.

Some have argued that standards efforts will eliminate conceptual heterogeneity, but this is not the case. First of all, although standards may evolve within specific application areas, they are likely to lag actual usage, coming only after years of work and changing only very slowly in response to new circumstances. Another problem with standards is that they apply only to specific application areas. In our modern information infrastructure, information processing must often cross application boundaries for which joint standards are not likely to exist (and arguably should not exist).

As an example of conceptual heterogeneity, consider a data integration setting in the area of domestic cookware. In our example, there are four data sources and three data consumers. The example is somewhat lengthy, but it has the merit of illustrating the most common types of heterogeneity encountered in practice and thereby conveying the complexity of the data integration problem.

One source in our example is the Carter database (Figure 1.2). The database contains a single table listing the cookware products manufactured by Carter together with various attributes of those products. Each row in the table represents a single product. The first column gives the unique identifier for the product; the second column specifies the type of product; the third column specifies the material; and the fourth column gives the manufacturer's suggested retail price (msrp) in US dollars.

| cookware | | | |
|---|---|---|---|
| id | type | material | price |
| c01 | skillet | aluminum | 50 |
| c02 | saucepan | aluminum | 40 |
| c03 | skillet | iron | 30 |
| c04 | saucepan | iron | 20 |

FIGURE 1.2: Carter database.

The Mirkwood database (Figure 1.3) lists the products manufactured by Mirkwood. In this case, there are multiple tables, one for each product attribute. The `kind` table specifies the type of each product. The `coated` table specifies those products that have nonstick coatings. The `price` table gives the msrp. Note that there is no material information. All of Mirkwood's products are made from either aluminum or stainless; and, because these materials have similar properties, the company has chosen not to provide information about the metal used in each product. Note that the `coated` table has only positive values; products without nonstick coatings are left unmentioned. The second column is there, in this case, simply to ensure that all of Mirkwood's tables are binary for the convenience of the database administrator.

| kind | | coated | | price | |
|---|---|---|---|---|---|
| **id** | **value** | **id** | **value** | **id** | **value** |
| m01 | skillet | m01 | yes | m01 | 60 |
| m02 | skillet | m02 | yes | m02 | 50 |
| m03 | saucepan | | | m03 | 40 |
| m04 | saucepan | | | m04 | 20 |

**FIGURE 1.3:** Mirkwood database.

The Marvel database (Figure 1.4) shows products listed by a company called Marvel, which provides data services for small companies that do not want to manage their own databases. In this

| kitchenware | | |
|---|---|---|
| **id** | **attribute** | **value** |
| r01 | maker | renfrew |
| r01 | type | skillet |
| r01 | material | aluminum |
| r01 | msrp | 50 |
| s03 | maker | superchef |
| s03 | type | skillet |
| s03 | material | stainless |
| s03 | coating | teflon |
| s03 | msrp | 30 |
| s04 | maker | superchef |
| s04 | type | skillet |
| s04 | material | iron |
| s04 | coating | ceramic |
| s04 | msrp | 20 |

**FIGURE 1.4:** Marvel database.

case, there is just one table, called `kitchenware`. However, there are multiple rows per product, one for each object, attribute, and value. The structure here is the sort that would be appropriate for representing triple-based data, as in Resource Description Format (RDF). The data here includes information about products made by Renfrew and Superchef, companies that do not have their own databases.

Our final source is a database maintained by the National Housewares Manufacturers Association (NHMA). In this case, there are three relations shown in Figure 1.5. The first relation, called `nonstick`, lists various materials and a Boolean indicating whether each material is nonstick. The second table lists cookware manufacturers and their nationalities. The third lists countries and the regions where they are located.

| nonstick | |
|---|---|
| id | value |
| ceramic | no |
| copper | no |
| teflon | yes |

| company | |
|---|---|
| id | nationality |
| carter | usa |
| mirkwood | uk |
| renfrew | canada |
| superchef | france |

| country | |
|---|---|
| id | area |
| canada | america |
| france | europe |
| uk | europe |
| usa | america |

FIGURE 1.5: NHMA database.

Like data providers, data consumers have relations and vocabularies they prefer. These relations and vocabularies can be different from each other as well as from the relations and vocabularies used by the data providers.

The first data consumer (Figure 1.6) in our example is a comparison shopping site, named Xanadu, that wants to provide its users with a database of cookware products made in the United

| catalog | | | |
|---|---|---|---|
| id | maker | type | price |
| c01 | carter | skillet | 50 |
| c02 | carter | saucepan | 40 |
| c03 | carter | skillet | 30 |
| c04 | carter | saucepan | 20 |
| r01 | renfrew | skillet | 50 |

FIGURE 1.6: Xanadu compilation.

States together with their manufacturers, types, and msrp values. Looking at the data sources described above, we can see that there are five products meeting the specifications. Based on the NHMA data, we know that only Carter and Renfrew are U.S. companies. The Carter database contains just four products, and the Marvel database lists just one product for Renfrew. The values for the various attributes can be found in these databases as well.

The second data consumer (Figure 1.7) is another comparison shopping site, named Yankee. In this case, the content is the same, but there are differences in vocabulary. Yankee uses `frypan` instead of `skillet` and `pot` instead of `saucepan`. Also, the msrp values are given in euros rather than dollars. In this case, there are five products, again taken from the Carter and Marvel databases.

| product | | | |
|---|---|---|---|
| **id** | **maker** | **type** | **price** |
| c01 | carter | frypan | 40 |
| c02 | carter | pot | 32 |
| c03 | carter | frypan | 24 |
| c04 | carter | pot | 16 |
| r01 | renfrew | frypan | 40 |

FIGURE 1.7: Yankee compilation.

The third data consumer (Figure 1.8) is a product review site, named Zebulon, featuring a table that lists products manufactured in Europe that are made from non-corrosible materials (e.g., aluminum, stainless, ceramic, and glass). The table includes information about the manufacturer of each product, the type, and the msrp of the product in dollars. In this case, there are just three products—two manufactured by Mirkwood and one manufactured by Superchef. Note that, in this case, we do not have detailed information about the materials for the two Mirkwood products,

| review | | | |
|---|---|---|---|
| **id** | **maker** | **type** | **price** |
| m01 | mirkwood | skillet | 60 |
| m02 | mirkwood | skillet | 50 |
| s03 | superchef | skillet | 30 |

FIGURE 1.8: Zebulon compilation.

since this information is not stored in any of the databases. The products are nonetheless included because we know, as mentioned above, that all Mirkwood products are made from either aluminum or stainless.

The goal of data integration is to create a broker capable of populating consumer tables like these, thereby saving the consumers the cost of entering the data manually. The examples here illustrate just some of the problems that must be overcome in order to create such a broker. In this book, we look at various ways of dealing with these problems and accomplishing this goal.

Note, however, that these are not the only problems that must be solved. The following are some common problems that we do not address in this book.

First of all, there is the problem of *inconsistency*. It is not uncommon for different databases to have conflicting data. In the absence of a technique for knowing which database to trust, a data broker must be able to manage multiple possible values without inappropriately combining those values to produce silly results. This is often called *paraconsistent reasoning*. The good news is that there are good techniques of this sort. Unfortunately, these techniques are beyond the scope of the book, and we do not describe them here.

Second, there is the problem of *irreconcilable naming*. Different databases may have different identifiers for the same objects. Sometimes this problem can be dealt with by using other data to relate entries to each other, a process called *entity resolution*. However, this is not always possible; there simply may not be enough information in the databases to decide which objects in one database correspond to which objects in the other databases. The good news here is that this problem can be mitigated or even eliminated by semantic web technologies of growing popularity, notably the use of *universal resource indicators* for objects and various sorts of *internet name servers* for registering and managing such names.

## 1.3 DIRECT MAPPING

Data changes rapidly, schemas and vocabularies change often but more slowly. This motivates an approach to brokering in which an administrator writes rules that allow the broker to convert questions written in terms of consumer schemas into questions written in terms of source schemas.

There are various approaches to doing this. In this section, we examine the conceptually simplest approach, called direct mapping, in which the administrator manually defines every relation in every consumer schema in terms of the relations in the source schemas.

In our examples here and throughout the book, we encode relationships between and among schemas as *rules* in a language called Datalog. In many cases, the rules are expressed in a simple version of Datalog called Basic Datalog; in other cases, rules are written in more elaborate versions, viz., Functional Datalog and Disjunctive Datalog. In the following paragraphs, we look at Basic

Datalog first, then Functional Datalog, and finally Disjunctive Datalog. The presentation here is casual; formal details are given in Chapter 2.

A *basic rule* is an expression of the form shown below. Here, $p, p_1, \ldots, p_k$ are names for relations, and the various $t_i$ and $t_{ij}$ are variables or constants. $p(t_1, \ldots, t_n)$ is called the conclusion of the rule, and the expressions $p_1(t_{11}, \ldots, t_{1n}), \ldots, p_k(t_{k1}, \ldots, t_{kn})$ are called conditions.

$$p(t_1, \ldots, t_n) \colon\!- p_1(t_{11}, \ldots, t_{1n}) \& \ldots \& p_k(t_{k1}, \ldots, t_{kn})$$

Semantically, a rule is something like reverse implication. It is a statement that the conclusion of the rule is true whenever the conditions are true (once constants have been substituted in a consistent way for all variables).

As an example of a basic rule, consider the expression shown below. This rule defines the grandparent relation in terms of the parent relation. One person is the grandparent of a second person if and only if the first person has a child and that child is the parent of the second person.

```
grandparent(X,Z)  :- parent(X,Y) & parent(Y,Z)
```

In the standard notation for Datalog, variables begin with upper case letters, and constants begin with lower case letters or digits; hence, all of the arguments here are variables.

Note that the same relation can appear as a conclusion in more than one rule. For example, parenthood can be defined in terms of fatherhood and motherhood as shown below.

```
parent(X,Y)  :- father(X,Y)
parent(X,Y)  :- mother(X,Y)
```

The need for existential rules arises when it is necessary to assert the existence of an object without knowing its identity. In Functional Datalog, we can refer to such an anonymous object using a *Skolem* term, i.e., an arbitrary function constant applied to the variables in the rule.

As examples of this, consider the two expressions shown below. Together, they characterize the parent relation in terms of the grandparent relation, i.e., the inverse of the definition given above. As mentioned above, one person is the grandparent of a second person if and only if the first person has a child and that child is the parent of the second person. The Skolem term `f(X,Z)` designates the intermediate child in any such relationship. We cannot use a constant here because we want the rule to work for all people. We cannot use a variable as that would say that the first person is the parent of everyone. The Skolem term in this case expresses the fact that such an intermediate person exists, and it relates the person to his or her parent and child.

```
parent(X,f(X,Z))  :- grandparent(X,Z)
parent(f(X,Z),Z)  :- grandparent(X,Z)
```

Skolem terms are similar to *null values* as used in relational databases. The main difference is that Skolem terms include variables to make explicit the dependence between each anonymous object and the objects on which it depends. As a result, we can safely equate syntactically identical Skolem terms (because they must refer to the same objects), whereas it is not safe to equate null values (since they may refer to different objects).

The need for disjunctive rules arises when there is more than one possible conclusion for a given set of conditions and it is not clear which conclusion holds. In Disjunctive Datalog, this situation is handled by writing multiple conclusions on the conclusion side of the rule.

The expression shown below is an example of a disjunctive rule. It states that, if one person is the parent of a second person, then that person is either the father or the mother of the first person.

```
father(X,Y) | mother(X,Y) :- parent(X,Y)
```

In this case, only one of the conclusions can follow, since a person cannot be both a father and a mother. In general, it is possible that both of the conclusions hold (though not in this case).

As an example of direct mapping using Datalog, consider the integration problem introduced in the last section. The following rules capture the relationship between the `catalog` relation and the relations in the source databases. Recall that Xanadu is intended to provide information about cookware products made in America.

```
catalog(Id,carter,Type,Msrp) :-
   cookware(Id,Type,Mat,Msrp)

catalog(Id,Mfr,Type,Msrp) :-
   kitchenware(Id,maker,Mfr) &
   company(Mfr,Country) &
   country(Country,america) &
   kitchenware(Id,type,Type) &
   kitchenware(Id,msrp,Msrp)
```

This first rule says that the Xanadu database should contain a datum of the form `catalog (Id,carter,Type,Msrp)` whenever there is a datum in the Carter database that matches `cookware(Id,Type,Mat,Msrp)`. Here, `carter` is a specific constant, and the symbols beginning with capital letters (e.g., `Id`, `Type`, `Mat`, and `Msrp`) are variables whose values range over constants that occur in the databases. There is no condition on nationality since we know that Carter is an American company.

The second rule says that the Xanadu database should also contain a datum of the form `catalog(Id,Mfr,Type,Msrp)` whenever (1) there is a datum in the Marvel database that matches `kitchenware(Id,maker,Mfr)`, (2) there is a datum in the NHMA database of the form `company(Mfr,Country)`, (3) there is a datum of the form `country(Country, america)`, (4) there is a datum in the Marvel database of the form `kitchenware(Id,type, Type)`, and (5) a datum `kitchenware(Id,msrp,Msrp)`.

Similar rules can be written for the product table. In this case, the rules are more complicated in order to deal with the change of vocabulary required for the Yankee database.

```
product(Id,carter,Type2,Msrp2) :-
    cookware(Id,Type1,Mat,Msrp1) &
    concordance(Type1,Type2) &
    exchange(euro,Rate) &
    times(Msrp1,Rate,Msrp2)

product(Id,Mfr,Type2,Msrp2) :-
    kitchenware(Id,maker,Mfr) &
    company(Mfr,Country) &
    country(Country,america) &
    kitchenware(Id,type,Type1) &
    concordance(Type1,Type2) &
    kitchenware(Id,msrp,Msrp1) &
    exchange(euro,Rate) &
    times(Msrp1,Rate,Msrp2)

concordance(roaster,pot)
concordance(skillet,frypan)
concordance(saucepan,pot)
concordance(wok,frypan)

exchange(euro,0.8)
exchange(loonie,1.2)
exchange(yen,1.0)
```

The first rule here says that the Yankee database should contain a datum of the form `product (Id,carter,Type2,Msrp2)` whenever there is a datum in the Carter database that matches `cookware(Id,Type1,Mat,Msrp1)`, where the value of `Type1` in the Carter vocabulary corresponds to the value of `Type2` in the Yankee vocabulary and the value of `Msrp1` in dollars corresponds to the value of `Msrp2` in euros. The second rule is analogous.

The concordance and exchange relations are included to facilitate the vocabulary translation. For each type in the Carter and Marvel vocabularies, concordance lists the corresponding word in the Yankee vocabulary. The exchange table provides exchange rates between dollars and other forms of currency. The times relation holds of all triples of numbers where the third number is the product of the first two numbers. times is a predefined relation in Datalog; hence, we do not provide a definition here.

Finally, we have rules for the review table. In this case, there is no change of vocabulary. However, there is complexity due to the requirements for inclusion of products in the Zebulon database. Recall that Zebulon is a compilation of all European products that are made from a non-corrosible materials.

```
review(Id,mirkwood,Type,Msrp) :-
   kind(Id,Type) &
   price(Id,Msrp)

review(Id,Mfr,Type,Msrp) :-
   kitchenware(Id,maker,Mfr) &
   company(Mfr,Country) &
   country(Country,europe) &
   kitchenware(Id,type,Type) &
   kitchenware(Id,material,Mat) &
   noncorrosible(Mat) &
   kitchenware(Id,msrp,Msrp)

noncorrosible(aluminum)
noncorrosible(stainless)
noncorrosible(ceramic)
noncorrosible(glass)
```

The first rule here says that the Zebulon database should contain a datum of the form review(Id,mirkwood,Type,Msrp) whenever there is a datum in the Mirkwood database that matches kind(Id,Type), and where there is a datum matching msrp(Id,Msrp). There is no condition on nationality since we know that Mirkwood is a European company. There is no condition on material since we know that all Mirkwood products are made from either aluminum or stainless, both non-corrosible materials. (See the description of Mirkwood in Section 1.2.) The second rule in this case includes conditions on the nationality of the manufacturer as well as a condition on materials.

Conceptually, direct mapping is the simplest approach to dealing with conceptual heterogeneity. In practice, however, it has a serious problem that limits its utility. Direct mapping deals with conceptual heterogeneity on a pairwise basis by defining the relations in the client's schema in terms of the relations in the source schemas. This leads to the so-called *n-squared problem*. The approach works fine when there are just a small number of clients and sources. However, it is not so good for large numbers of clients and sources. What happens when there are thousands of clients and sources? In the worst case, we need definitions for every client schema in terms of every source schema, making for millions of combinations.

In the next two sections, we look at two different approaches to solving this problem. Both approaches solve the *n-squared* problem through the use of a *master schema*. All client schemas and all source schemas are mapped to this master schema. The two approaches differ in terms of how the definitions are written. In source-based integration, the relations in the master schema are defined in terms of the relations in the source schemas. In the model-centric approach, the definitions go the other way: the relations in the source schema are defined in terms of the relations in the master schema. As we shall see, the model-centric approach has all of the benefits of the source-centric approach but makes things easier for database administrators.

## 1.4   SOURCE-BASED INTEGRATION

In *source-based integration* (Figure 1.9), each client schema is defined in terms of a common master schema, and each relation in this master schema is defined in terms of the relations in the source schemas. This approach is also sometimes known as *global as view integration*, or GAV integration, since the relations in the global/master schema are defined as views of the sources.

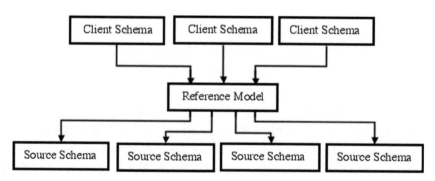

FIGURE 1.9: Source-based integration.

Since there is only one conversion for each client and each source, the cost of this approach is linear in the number of schemas to be integrated—seven schemas, seven rule sets.

As an example of this approach to integration, once again consider the problem described in Section 1.2. Let us assume that we have a master schema based on individual tables for various product attributes. The maker table lists products with their manufacturers. The type table lists products with their types. The material table lists products with their materials. The coating table lists products with their coatings (if any). The price table lists products with their msrp values in dollars. We also include the nonstick and company and country tables from the NHMA database.

Using this schema, we can define the various consumer schemas as shown below.

```
catalog(Id,Mfr,Type,Msrp) :-
  maker(Id,Mfr) &
  company(Mfr,Country) &
  country(Country,america) &
  type(Id,Type) &
  msrp(Id,Msrp)

product(Id,Mfr,Type2,Msrp2) :-
  maker(Id,Mfr) &
  company(Mfr,Country) &
  country(Country,america) &
  type(Id,Type1) &
  concordance(Type1,Type2) &
  msrp(Id,Msrp1) &
  exchange(euro,Rate) &
  times(Msrp1,Rate,Msrp2)

review(Id,Mfr,Type,Msrp) :-
  maker(Id,Mfr) &
  company(Mfr,Country) &
  country(Country,europe) &
  type(Id,Type) &
  material(Id,Mat) &
  noncorrosible(Mat) &
  msrp(Id,Msrp)
```

In turn, we can define the relations in our master schema in terms of the relations in the various data sources. If we restrict ourselves to Basic Datalog, the following is the best we can do.

```
maker(Id,carter)  :- cookware(Id,Type,Mat,Msrp)
maker(Id,mirkwood) :- kind(Id,Type)
maker(Id,Mfr)  :- kitchenware(Id,Maker,Mfr)

type(Id,Type)  :- cookware(Id,Type,Mat,Msrp)
type(Id,Type)  :- kind(Id,Type)
type(Id,Type)  :- kitchenware(Id,type,Type)

material(Id,Mat)  :- cookware(Id,Type,Mat,Msrp)
material(Id,Mat)  :- kitchenware(Id,material,Mat)

coating(Id,Ctg)  :- kitchenware(Id,coating,Ctg)

msrp(Id,Msrp)  :- cookware(Id,Type,Mat,Msrp)
msrp(Id,Msrp)  :- msrp(Id,Msrp)
msrp(Id,Msrp)  :- kitchenware(Id,msrp,Msrp)
```

Restricting ourselves to Basic Datalog has the merit that all of the definitions are simple and they can all be translated directly into languages like SQL. The downside is that in doing things this way, we lose answers. In this case, notice that there are no rules connecting the material or coating relations to the relations in the Mirkwood database, meaning that no Mirkwood products would be found in cases where material or coating are specified. This is problematic because we actually do have enough information about the materials and coatings used in Mirkwood's products to produce at least partial answers.

This problem can be remedied by writing rules in one of the extensions to Datalog. The coating of Mirkwood products can be characterized using Functional Datalog.

```
coating(Id,f(Id))  :- coated(Id,yes)
```

Using Disjunctive Datalog, it is possible to characterize the material relation in terms of price as shown below.

```
material(Id,aluminum) | material(Id,stainless) :-
    price(Id,Y)
```

Using these additional rules, a data broker would be able to find answers that would not otherwise be possible. For example, if the consumer were interested in metallic products with coatings, as in the case of Zebulon, a broker would still be able to provide answers.

The downside of using extended versions of Datalog is that the rules are more complex, and there are fewer people with the ability to write such rules. Moreover, there is no direct translation to popular languages like SQL and SPARQL. Although these languages contain unions, the semantics of the languages would lead to no answers in cases like this.

## 1.5    MODEL-CENTRIC INTEGRATION

The good news is that it is possible to have the benefits of Functional and Disjunctive Datalog without this extra complexity. The answer is to provide users with a master schema and allow database administrators to write Basic Datalog rules defining their relations in terms of the relations in this master schema. This is *model-centric integration* (Figure 1.10). It is also called *local as view (LAV) integration*, since the relations in the sources are defined as views of the relations in the master schema. The following diagram illustrates this idea.

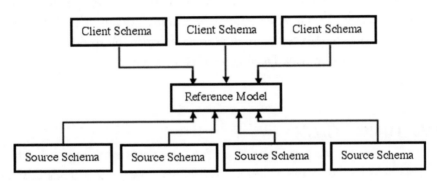

FIGURE 1.10: Model-centric integration.

A well-designed master schema is more detailed than any of the schemas it is used to integrate. The importance of this is that it is possible for administrators to define client and source relations in terms of reference relations using just Basic Datalog (or even SQL). They need not write existential or disjunctive rules at all!

```
cookware(Id,Type,Mat,Msrp) :-
  maker(Id,carter) &
  type(Id,Type) &
  material(Id,Mat) &
  msrp(Id,Msrp)
```

```
kind(Id,Type) :- maker(Id,mirkwood) & type(Id,Type)
coated(Id,yes) :-
  maker(Id,mirkwood) &
  coating(Id,Y) &
  nonstick(Y,yes)
price(Id,Msrp) :- maker(Id,mirkwood) & msrp(Id,Msrp)

kitchenware(Id,maker,Company) :- maker(Id,Company)
kitchenware(Id,type,Type) :- type(Id,Type)
kitchenware(Id,material,Mat) :- material(Id,Mat)
kitchenware(Id,coating,Ctg) :- coating(Id,Ctg)
kitchenware(Id,msrp,Msrp) :- msrp(Id,Msrp)
```

In order for data broker to implement this model-centric approach, queries must be written in terms of source relations. This is called *query folding*. In this case, the aim of query folding is to produce the rules shown in the last section from the rules shown in this section. In this way, the data administrators can write rules in Basic Datalog, yet we get the power of Functional and Disjunctive Datalogs.

## 1.6   READING GUIDE

In this book, we concentrate on data integration using the model-centric approach. This approach has the advantages described in the preceding section. Moreover, the methods used for model-centric integration are a superset of the methods needed for direct mapping and source-based integration; hence, in a sense, we are covering all three approaches.

Query processing in model-centric integration consists of three phases—query folding, query planning, and plan execution.

Upon receipt of a query, a data broker first uses query folding to rewrite the query in terms of the source relations, resulting in a *retrievable query*. In this book, we concentrate on a method of query folding called the Inverse Method. The details of this method are described in Chapter 3.

Given a retrievable query, a data broker then decides how to execute the retrievable query. The main issue here is how to divide up the query into individual steps and what order to use in executing the steps. Moreover, in the presence of multiple sources of the same relation, it must decide which sources to use. The result of query planning is an *executable query plan*. We discuss various approaches to query planning in Chapter 4.

Finally, given an executable query plan, the broker executes the plan—sending queries to sources, collecting the results, combining and transforming them as appropriate, and sending the results back to the user. This phase uses traditional methods for execution of distributed query plans, and we do not discuss it further here.

Note that effective model-centric data integration requires a good master schema. Prior to deploying a data integration system, the designer of a data broker must select a master schema, and the designer must occasionally revise this schema to deal with new complexities in the data integration environment. In some cases, the designer can rely on automated methods for master schema design and redesign. In Chapter 5, we present some issues in master schema design and suggest some methods for automatic formulation and reformulation.

.   .   .   .

CHAPTER 2

# Basic Concepts

## 2.1 RELATIONAL DATABASES

The fundamental building blocks of databases are entities and relations. *Entities* represent objects presumed or hypothesized to exist in the application area of the database. *Relations* represent properties of those objects or relationships among them.

In our examples here, we refer to entities and relations using strings of letters, digits, and a few non-alphanumeric characters (e.g., "_"). For reasons described below, we prohibit strings beginning with upper case letters; all other combinations are acceptable. Examples include `a`, `b`, `123`, `comp225`, `helen_heavenly`, and `cAmE1`.

The set of all entities that can be used in a database is called the *domain* of the database. Database domains can be finite or infinite. Most, but not all, database domains include arbitrary integers as subsets, and hence they are infinite.

The set of all relations in a database is called the *signature* of the database. Signatures are always finite.

The *arity* of a relation is the number of objects involved in any instance of that relation. Consider, for example, the teaching schedule in a university database. Every instance of this relation involves two objects, i.e., a faculty member and a course the faculty member teaches; therefore, it has arity 2. Arity is an inherent property of a relation and never changes.

A database *schema* consists of a domain, a signature, and an assignment of arities for each of the relations in the signature. Our definition here departs slightly from that used in many database texts. Usually, a schema does not include a restriction on the entities used in relations, whereas we have added in a domain of possible entities here. This addition simplifies our definitions but does not change any fundamental results.

In the *relational data model*, the *extension* of an $n$-ary relation in a database is a set of $n$-tuples from the domain of the database. The *cardinality* of a relation in an extension is the number of tuples of objects that satisfy the relation in a particular state. Unlike arity, the cardinality of a relation can change from one extension to another.

In the database literature, it is common to present the extension of a relation as a two-dimensional table. The number of columns in the table corresponds to the arity of the relation and the number of rows corresponds to the cardinality of the extension.

---

*An interactive version of this book is available that includes implementations of algorithms used in demonstrations and exercises. See page xi for more information.*

For example, the extension of the `teaches` relation is a set of pairs of faculty members and courses, one pair for each faculty member and each course that the faculty member teaches. If we consider a state of the world in which there are six pairs of faculty members and courses that they teach, we can visualize this extension as a table with two columns and six rows, as shown in Figure 2.1. Here, the extension has arity 2 and cardinality 6.

| teaches | |
| --- | --- |
| **faculty** | **course** |
| donna_daring | arch101 |
| bill_boring | comp101 |
| cathy_careful | comp225 |
| gary_grump | comp235 |
| oren_overbearing | comp257 |
| helen_heavenly | comp310 |

FIGURE 2.1: A binary relation.

The `student` relation is a property of a person in and of itself, not with respect to other people or other objects. Since there is just one object involved in any instance of the relation, the table has just one column. By contrast, there are six rows, one row for each student. In this case, the extension has arity 1 and cardinality 6.

| student |
| --- |
| aaron_aardvark |
| belinda_bat |
| calvin_carp |
| george_giraffe |
| kitty_kat |
| sally_squirrel |

FIGURE 2.2: A unary relation.

The `gradesheet` relation is a relation among students and courses and grades; thus, the table requires three columns, as shown in Figure 2.3. In this case, we have an extension with arity 3 and cardinality 4.

| gradesheet | | |
|---|---|---|
| **student** | **course** | **grade** |
| aaron_aardvark | arch101 | a |
| aaron_aardvark | comp101 | a |
| calvin_carp | arch101 | b |
| sally_squirrel | comp101 | a |

**FIGURE 2.3:** A ternary relation.

This tabular representation for relations makes clear that, for a finite domain, there is an upper bound on the number of possible extensions for an $n$-ary relation. In particular, for a universe of discourse of size $b$, there are $b^n$ distinct $n$-tuples. Every extension of an $n$-ary relation is a subset of these $b^n$ tuples. Therefore, the extension of an $n$-ary relation must be one of at most $2^{b^n}$ possible sets.

## 2.2  SENTENTIAL DATABASES

While the relational data model is an appealing way to think about data, the *sentential data model* is more useful for our purposes here. Everything is the same as in the relational data model up to the definition of an extension. In the sentential data model, we encode each instance of a relation in the form of a sentence consisting of the relation and the entities involved in the instance, and we define a database extension as a set of such sentences.

More precisely, given a database, we define a *datum* to be a structure consisting of an $n$-ary relation from the signature and $n$ entities from the domain. On occasion, we call a datum a *proposition*. In what follows, we write data using traditional mathematical notation. For example, if r is a binary relation and a and b are entities, then r(a,b) is a datum/proposition.

Consider the data shown in Figure 2.1. We can encode this in sentential representation as shown below. There is one sentence for each row in the original table. Each sentence here has the same relation. The relation is included so that we can combine these sentences with sentences representing other relations.

```
teaches(donna_daring,arch101)
teaches(bill_boring,comp101)
teaches(cathy_careful,comp225)
teaches(gary_grump,comp235)
teaches(oren_overbearing,comp257)
teaches(helen_heavenly,comp310)
```

The sentences corresponding to the data in Figure 2.2 are shown below; `student` is a unary relation, thus the sentences in this case all have one argument.

```
student(aaron_aardvark)
student(Belinda_bat)
student(calvin_carp)
student(george_giraffe)
student(kitty_kat)
student(sally_squirrel)
```

Finally, we have the sentences for the data in Figure 2.3. Here, each sentence has three arguments, since the relation `gradesheet` has arity 3.

```
gradesheet(aaron_aardvark,arch101,a)
gradesheet(aaron_aardvark,comp101,a)
gradesheet(calvin_carp,arch101,b)
gradesheet(sally_squirrel,comp101,a)
```

Since there are no column headings in such sentences in this presentation, as there are in tables, the order of arguments is important. Given the intended meaning of the `gradesheet` relation, in our example, the first argument denotes the student, the second the course, and the third the grade.

The *propositional base* for a database schema is the set of all propositions that can be formed from the relations and the entities in the database schema. For a schema with entities a and b and relations p and q where p has arity 1 and q has arity 2, the propositional base is {p(a), p(b), q(a,a), q(a,b), q(b,a), q(b,b)}.

A *sentential database* is a database in which the state is expressed in sentential representation. An *extension* of a sentential database is a finite subset of its propositional base.

## 2.3 DATALOG PROGRAMS

Datalog programs are built up from three disjoint classes of components, i.e., *relations*, *entities*, and *variables*. In our examples here, we denote entities and relations as described above; and we write variables as strings of letters, digits, and special characters beginning with an upper case letter, e.g., X, Y, Z, Age, F1, F2, etc. A *term* is either an entity or a variable.

An *atom* is an expression formed from an *n*-ary relation and *n* terms. As with database sentences, we write Datalog atoms in traditional mathematical notation: the relation followed by its *arguments* enclosed in parentheses and separated by commas. For example, if r is a binary relation, if a and b are entities, and if Y is a variable, then r(a,b) and r(b,a) are atoms, as before, but so are r(a,Y), r(Y,a), and r(Y,Y).

A *literal* is either an atom or a negation of an atom. An atom is called a *positive* literal. The negation of an atom is called a *negative* literal. In what follows, we write negative literals using the negation sign ~. For example, if r(a,b) is an atom, then ~r(a,b) denotes the negation of this atom.

A *rule* is an expression consisting of a distinguished atom, called the *head*, and zero or more literals, together called the *body*. In what follows, we write rules as in the example shown below. Here, q(X,Y) is the head, and the other literals constitute the body.

$$q(a,Y) \ :- \ p(b,Y) \ \& \ \sim r(Y,d)$$

A *Datalog program* is a finite set of atoms and rules of this form. In order to simplify our definitions and analysis, we occasionally talk about infinite sets of rules. While these sets are useful for purposes of analysis, they are not themselves Datalog programs.

An expression in Datalog is said to be *ground* if and only if it contains no variables.

A rule in a Datalog program is *safe* if and only if every variable that appears in the head or in any negative literal in the body also appears in at least one positive literal in the body. A Datalog program is safe if and only if every rule in the program is safe.

The *dependency graph* for a Datalog program is a directed graph in which the nodes are the relations in the program and in which there is an arc from one node to another if and only if the former node appears in the body of a rule in which the latter node appears in the head. A program is *recursive* if and only if there is a cycle in the dependency graph.

A negation in a Datalog program is said to be *stratified* if and only if there is no recursive cycle in the dependency graph involving the negation. A Datalog program is stratified if and only if there are no unstratified negations.

In this book, we concentrate exclusively on Datalog programs that are both safe and stratified. While it is possible to extend the results here to other programs, such extensions are beyond the scope of this work.

The *propositional base* for a Datalog program is the set of all atoms that can be formed from the entities in the program's schema. Said another way, it is the set of all sentences of the form $r(t_1, \ldots, t_n)$, where $r$ is an $n$-ary relation and $t_1, \ldots, t_n$ are entities.

An *instance* of a rule in a Datalog program is a rule in which all variables have been consistently replaced by terms from the program's domain. *Consistent replacement* means that if one occurrence of a variable is replaced by a given term, then all occurrences of that variable are replaced by the same term.

An *interpretation* for a Datalog program is an arbitrary subset of the propositional base for the program. A *model* of a Datalog program is an interpretation that satisfies the program (as defined below).

An interpretation $D$ *satisfies* a Datalog program $P$ if and only if $D$ satisfies every ground instance of every sentence in $P$. The notion of satisfaction is defined recursively. An interpretation $D$ satisfies a ground atom $p$ if and only if $p$ is in $D$. $D$ satisfies a ground negation $\sim p$ if and only if $p$ is *not* in $D$. $D$ satisfies a ground rule $p$ :- $p_1$ & ... & $p_n$ if and only if $D$ satisfies $p$ whenever it satisfies $p_1, \ldots, p_n$.

In general, a Datalog program can have more than one model, which means that there can be more than one way to satisfy the rules in the program. In order to eliminate ambiguity, we adopt the minimal model approach to Datalog program semantics, i.e., we define the meaning of a safe and stratified Datalog program to be its minimal model.

A model $D$ of a Datalog program $P$ is *minimal* if and only if no proper subset of $D$ is a model for $P$. A Datalog program that does not contain any negations has one and only one minimal model. A Datalog program with negation may have more than one minimal model; however, if the program is stratified, then once again there is only one minimal model.

## 2.4    OPEN DATALOG PROGRAMS

Datalog programs as just defined are *closed* in that they fix the meaning of all relations in the program. In open Datalog programs, some of the relations (the inputs) are undefined, and other relations (the outputs) are defined in terms of these. The same program can be used with different input relations, yielding different output relations in each case.

Formally, an *open program* is a Datalog program together with a partition of the relation constants into two types: *base relations* (also called *input relations*) and *view relations* (also called *output relations*). View relations can appear anywhere in the program, but base relations can appear only in the bodies of rules, not in their heads.

The *input base* for an open Datalog program is the set of all atoms that can be formed from the base relations of the program and the entities in the program's domain. An *input model* is an arbitrary subset of its input base.

The *output base* for an open Datalog program is the set of all atoms that can be formed from the view relations of the program and the entities in the program's domain. An *output model* is an arbitrary subset of its output base.

Given an open Datalog program $P$ and an input model $D$, we define the overall model corresponding to $D$ to be the minimal model of $P \cup D$. The output model corresponding to $D$ is the intersection of the overall model with the program's output base.

Finally, we define the meaning of an open Datalog program to be a function that maps each input model for the program into the corresponding output model.

As an example, consider the simple open Datalog program shown below. The universe of discourse is {a,b,c}, p and q are base relations, and r is a view relation:

$$r(X,Y) :- p(X,Y) \ \& \ \sim q(Y)$$

In this case, there are $2^{12}$ input models, and for each there is a unique output model. The table below shows the output model for two of the possible input models.

| INPUT MODEL | OUTPUT MODEL |
|---|---|
| {p(a,c); q(b)} | {r(a,c)} |
| {p(a,c); p(b,c)} | {r(a,c); r(b,c)} |

## 2.5 DATABASE QUERIES

A *query* for a database $D$ is an open Datalog program with the following restrictions. (1) The program has a distinguished output relation (which, by convention, we call query). (2) The input relations of the program are all relations in $D$'s signature and have the same arity as in $D$'s schema.

The value of a query $Q$ on a database instance $D$, written $Q(D)$, is the set of all atoms in the minimal model of $Q \cup D$ that mention the query relation.

Query $Q_1$ is *contained in* query $Q_2$ (written $Q_1 \leq Q_2$) if and only if for every database instance $D$, $Q_1(D) \subseteq Q_2(D)$. Query $Q_1$ is *equivalent* to $Q_2$ if and only if $Q_1 \leq Q_2$ and $Q_2 \leq Q_1$.

The problem of determining whether a Datalog program $Q$ is contained in another Datalog program $Q$ is, in general, undecidable.

The problem becomes decidable if $Q$ contains only positive literals and is not recursive. The following is a decision procedure for this case. First, replace all distinct variables in $Q$ by distinct entities, yielding Q. Consider the database $D$ that consists of the literals in the bodies of all of these instantiated rules. $D$ is called the *canonical database* for $Q$. Evaluate $Q'$ on $D$. $Q$ is contained in $Q'$ if and only if the instantiated heads of all rules in $Q$ are contained in $Q'(D)$.

The expansion of a relation $r$ defined in a program $P$ (written $r^\infty$) is the set of rules obtained by substituting the definitions in $P$ for the relations in the bodies of all rules with $r$ in the head. Variables occurring in the bodies of definitions that do not appear in the head are replaced by new variables in the expansion.

## 2.6   DATABASE CONSTRAINTS

In our development thus far, we have assumed that the extension of an *n*-ary relation may be any set of *n*-tuples from the domain. This is rarely the case. Often, there are *constraints* that limit the set of possibilities. For example, a course cannot be its own prerequisite. In some cases, constraints involve multiple relations. For example, only students should appear in the first column of the grade relation; in other words, if an entity appears in the first column of the `grade` relation, it must also appear as an entry in the `student` relation.

Database management systems can use such constraints in a variety of ways. They can be used to optimize the processing of queries. They can also be used to check that updates do not lead to unacceptable extensions.

In many database texts, constraints are written in direct form by writing rules that say, in effect, if certain things are true in an extension, then other things must also be true. The *inclusion dependency* mentioned above is an example; if an entity appears in the first column of the `grade` relation, it must also appear as an entry in the `student` relation.

In what follows, we use a slightly less direct approach: we encode limitations by writing rules that say when a database is *not* well formed. For example, if an entity appears in the first column of the `grade` relation and does not appear in the `student` relation, then the extension is not well formed.

One merit of this approach is that we can use Datalog to write such rules. We simply invent a new 0-ary relation, here called `error`, and define it to be true in any extension that does not satisfy our constraints.

This approach works particularly well for consistency constraints like the one stating that a course cannot be its own prerequisite.

```
error :- prerequisite(X,X)
```

Using this technique, we can write the constraint on `grade` as shown below. There is an error if an entity is in the first column of the `grade` relation and it does not occur in the `student` relation:

```
error :- grade(X,Y,Z), ~student(X)
```

We can even write the *existential (tuple-generating) constraint* that every faculty member teaches at least one course, although, in this case, we must define a helper relation `teacher`.

```
error :- faculty(X), ~teacher(X)
teacher(X) :- teaches(X,Y)
```

In fact, using this approach, we can define constraints of all sorts. In order for the direct approach to work, it is necessary to introduce explicit *quantifiers* into rules. While this presents no theoretical difficulty, it means that we need to extend all of our results and methods to handle such constructs, whereas using the Datalog programming approach, we need nothing more than the theorems and algorithms described in the preceding sections.

## 2.7 PARTIAL DATALOG PROGRAMS

The rules in Datalog programs provide complete definitions of view relations in terms of base relations. In data integration, complete definitions are not always possible. On the other hand, we may know something about the extension of the view relation and it is desirable to capture this partial information.

In order to express what we do know about views, we can extend Datalog in various ways. In what follows, we consider two particular extensions: the introduction of *functional terms* and the introduction of *disjunctions*.

Consider, for example, the relation of parenthood and the relation of grandparenthood. While it is possible to define grandparenthood in terms of parenthood, it is not possible to define parenthood in terms of grandparenthood. On the other hand, if we were given an instance of the grandparenthood relation, we could say *something* about the parenthood relation. For example, if a person *a* is a grandparent of person *c*, we know that there is a person *b* such that *a* is the parent of *b* and *b* is the parent of *a*; we just do not know the identity of *b*.

Consider the kinship relations father and mother and parent. While it is possible to define parent in terms of father and mother, it is not possible to define father or mother in terms of parent. On the other hand, if we know what the parent relation holds of person *a* and person *b*, then we can know whether *a* is the father of *b* or *a* is the mother of *b*.

## 2.8 FUNCTIONAL DATALOG PROGRAMS

The main difference between Functional Datalog and Basic Datalog is the availability of functional terms to refer to entities in the domain of a database.

A *functional term* is an expression consisting of an *n*-ary function and *n* terms. In what follows, we write functional terms in traditional mathematical notation: the function followed by its *arguments* enclosed in parentheses and separated by commas. Also, if f is a function constant, a is an object constant, and Y is a variable, then f(a,Y) is a term. For example, if f(a,Y) is a functional term, then so is f(f(a,Y),Y).

In Functional Datalog, the *enhanced domain* for a program is the set of all entities and all functional terms that can be formed from the functions and entities in the program. Note that in the presence of functions, the enhanced domain is infinite. The propositional base is the set of all propositions that can be formed from the relations in the program and the objects in the enhanced domain.

An *interpretation* for a Functional Datalog program is an arbitrary subset of the propositional base for the program. A *model* of a Datalog program is an interpretation that satisfies the program (as defined above).

Note that, unlike Basic Datalog programs, Functional Datalog programs are not guaranteed to terminate since they can generate terms with arbitrarily nested functional terms. As an example, consider the following logic program.

$$r(X) \ :- \ p(X)$$
$$r(f(X)) \ :- \ r(X)$$

The result in this case may contain infinitely many terms. If the extension of the p relation contains the constant a, then the result includes a, f(a), f(f(a)), etc.

Fortunately, in many cases, termination can be assured, e.g., when the programs are not recursive or when all recursions are bounded in one way or another.

## 2.9   DISJUNCTIVE DATALOG PROGRAMS

A *disjunctive Horn rule* is a sentence of the following form, where every $q_i$ is an atom and every $p_i$ is a literal.

$$q_1 \ | \ \cdots \ | \ q_n \ :- \ p_1 \ \& \ \cdots \ \& \ p_m$$

Every variable in the head of a rule must also occur in the body of the rule. A *Horn rule* is a disjunctive Horn rule where the head consists of just one disjunct. A *disjunctive program* is a set of function-free rules.

A *conjunctive program* is a single non-recursive function-free Horn rule. A *positive program* is a set of conjunctive queries with the same relation in the head.

## 2.10   ENHANCED DATALOG PROGRAMS

On occasion, it is useful to write programs utilizing well-known relationships, such as arithmetic functions (e.g., +, *, -, /), string functions (e.g., concatenation), comparison operators (e.g., < and >), and equality (=). The definitions for most of these relations would require tables of infinite size. For this reason, we usually include these relations as part of the language itself, leading to *Enhanced Datalog*, *Enhanced Functional Datalog*, or *Enhanced Disjunctive Datalog*.

If the bodies of rules are allowed to contain these predefined relations, there are syntactic restrictions. For example, if rules contain literals with comparison operators, then every variable that occurs in such literals must appear in at least once in a positive literal in the body.

Even with this restriction, the presence of predefined relations can significantly affect the computational properties of the resulting programs. In the following chapters, we talk about these effects as we use various enhancements.

CHAPTER 3

# Query Folding

## 3.1    INTRODUCTION

Over the years, researchers have proposed several different methods for query folding, of which the best known are the Bucket Method, the Unijoin Algorithm, and the Inverse Method. In this book and in this chapter, we concentrate on the Inverse Method, as it is the most general and, by some measures, the most efficient.

The next section gives a crisp characterization of the query folding problem. The sections thereafter define the Inverse Method. We then compare it to the Bucket Method and the Unijoin Method. Finally, we discuss the decidability and complexity of the query folding problem, in general.

## 3.2    PROBLEM DEFINITION

The inputs to query folding are (1) a definition for a query relation in terms of a base schema and (2) definitions for the available source relations expressed in terms of the same base schema. The output is a definition for the query relation expressed entirely in terms of the available source relations. The set of rules in this definition is hereinafter called a *retrievable query plan*.

Ideally, the output of query folding would be a retrievable query whose expansion is *equivalent* to the original query. As described in Chapter 2, this means that, for every instance of the base relations, the answer to the original query is the same as the answer to the retrievable query.

Consider the Datalog query shown below, where edge and black are base relations. If edge is the connectivity relation for a graph and black is a coloring relation on the nodes of the graph, then the program defines the endpoints of paths of length two with a black internal node.

```
query(X,Z) :- edge(X,Y) & edge(Y,Z) & black(Y)
```

Now, suppose the two views defined below are source relations. View v1 records edges with black start nodes, and view v2 records edges with black end nodes.

```
v1(X,Y) :- edge(X,Y) & black(X)
v2(X,Y) :- edge(X,Y) & black(Y)
```

*An interactive version of this book is available that includes implementations of algorithms used in demonstrations and exercises. See page xi for more information.*

In this case, it is possible to produce a retrievable query that is equivalent to the original query. See below. The equivalence in this case can be easily seen by expanding the definitions for the source relations.

```
query(X,Z)  :-  v2(X,Y) & v1(Y,Z)
```

Unfortunately, it is not always possible to find an equivalent retrievable query. In our example, suppose that only v1 were available. Then there would not be an equivalent retrievable query.

The alternative to finding an equivalent retrievable query is finding one whose expansion approximates the original query. Most users are willing to accept some incompleteness in the result but are less willing to accept incorrect answers. The upshot is that most work on query folding is concerned with retrievable queries that are contained in the original queries.

Of all retrievable queries contained in an original query, maximal retrievable queries are the most interesting. Given a query $Q$, view definitions $V$, and a language $L$, a *maximal retrievable query* is one that contains every retrievable query for $Q$ and $V$ that can be expressed in $L$ and that is itself contained in $Q$.

As an example, consider the query and views defined above. Suppose that only v1 is available. Assume that only conjunctive views are possible. Then the plan shown below is a maximal retrievable query.

```
query(X,Z)  :-  v1(X,Y) & v1(Y,Z)
```

This query defines all paths of length two in which the start node and the intermediate node are both black. The expansion of the plan is contained in the original query. However, it leaves out paths in which the first node is not black; consequently, it is not equivalent to the original query.

In some cases, there may be retrievable queries but no maximal retrievable query that can be expressed within the chosen language. As a simple example of this, consider the query and views shown below.

```
query(X)  :-  m(x)
v1(X)  :-  m(X) & p(X)
v2(X)  :-  m(X) & q(X)
```

If we allow disjunctive queries, then the following is a maximal retrievable query for this case.

```
query(X)  :-  v1(X)
query(X)  :-  v2(X)
```

If we allow only conjunctive definitions, then this query is not permitted, as it is disjunctive. However, each of the two rules, taken alone, defines a different retrievable query. Unfortunately, there is no maximal retrievable query in this case, since neither one of the two retrievable queries contains the other.

## 3.3 INVERSE METHOD

The key step in the Inverse Method is the *inversion* of the definitions for all source relations, resulting in a set of rules defining the base relations in terms of the source relations. These rules are then added to the rules defining the query relations in terms of the base relations, resulting in an open Datalog program defining the query relations in terms of the source relations. This program is then simplified and optimized in various ways to form the final query plan.

As a simple example of the method in operation, consider the following data integration problem. There are three source relations. The first relation consists of pairs of cities between which Southwest Airlines (sw) has nonstop flights. The second relation contains similar information for United Airlines (ua). The third relation is the set of every triple consisting of an airline and cities that are connected by flights offered by that airline via exactly one stopover.

There is a single ternary base relation flight, which stores triples of airline, origin, and destination for all nonstop flights. For example, flight(sw,tus,sfo) means that Southwest Airlines offers a nonstop flight from Tucson to San Francisco. Using this base relation, we can define the source relations as shown below.

```
v1(F,T)   :- flight(sw,F,T)
v2(F,T)   :- flight(ua,F,T)
v3(C,F,T) :- flight(C,F,Z) & flight(C,Z,T)
```

In this case, the views can be inverted as shown below. Note the presence of the functional term g(C,F,T) in the third and fourth rules. This is called a Skolem term. Its occurrence in the third and fourth rules below captures the fact that, for every triple in the v3 relation, there are corresponding triples in the flight relation.

```
flight(sw,F,T)      :- v1(F,T)
flight(ua,F,T)      :- v2(F,T)
flight(C,F,g(C,F,T)) :- v3(C,F,T)
flight(C,g(C,F,T),T) :- v3(C,F,T)
```

Now, assume that the user wants a list of airlines that fly from Tucson (tus) to San Francisco (sfo) with at most one stopover. This query can be expressed as follows.

```
query(C)  :- flight(C,tus,sfo)
query(C)  :- flight(C,tus,Z) & flight(C,Z,sfo)
```

We form a retrievable query by combining these rules with the inverse rules for the source relations. By unwinding the definitions in this plan and simplifying, it is possible to produce the following retrievable query.

```
query(sw)  :-  v1(tus,sfo)
query(ua)  :-  v2(tus,sfo)
query(C)   :-  v3(C,F,T)
```

The difficulty of finding a retrievable version of a query (whether equivalent or maximally contained) increases with the logical complexity of the definitions for the source relations. In what follows, we deal with the case of conjunctively defined source relations first, followed by disjunctive source relations.

## 3.4    CONJUNCTIVE SOURCE DEFINITIONS

The simplest case of query folding occurs when all source relations are defined as conjunctive views of the base relations. The Inverse Method in this case is particularly simple.

In what follows, we assume that that every source relation $v$ is defined as a conjunctive view of base relations $u_1, \ldots, u_k$, as shown below.

$$v(\overline{X})  \; :- \; u_1(\overline{X}_1) \; \& \; \ldots \; \& \; u_k(\overline{X}_k)$$

The inverse in this case is the set of all rules having the original head [i.e., $v(\overline{X})$] as body and having each of the literals in the body as head. The variables in the original head remain unchanged. Every variable that appears in the original body but not in the head is consistently replaced by a term of the form $f(\overline{X})$, where $f$ is a new function constant chosen anew for every such variable. The inverse $V^-$ of a set of rules is the union of the inverses for each of the rules in the set.

As an example, once again consider the third source relation in the last section.

```
v3(C,F,T)  :-  flight(C,F,Z) & flight(C,Z,T)
```

The definition of the third relation has two conjuncts and, therefore, leads to two inverse rules.

```
flight(C,F,g(C,F,T))  :-  v3(C,F,T)
flight(C,g(C,F,T),T)  :-  v3(C,F,T)
```

Note that the inverse of a set of conjunctive view definitions is a Functional Datalog program but not necessarily a Basic Datalog program. This is worrisome because, as described in the preceding chapter, Functional Datalog programs are not guaranteed to terminate.

The interesting thing about the Inverse Method in the case of conjunctively defined views is that the result is better behaved. The construction produces logic programs whose bottom–up evaluation always terminates. The function constants are introduced only in inverse rules; and, because the inverse rules are not recursive, no terms with nested function constants are ever produced.

**Theorem:** For every open Datalog program $P$ and every set of conjunctively defined views $V$ and every database $D$, the logic program $P \cup V^- \cup D$ has a unique finite minimal model. Furthermore, evaluation is guaranteed to terminate and produce this model.

**Proof:** $P$ may be recursive but does not introduce function constants. $V^-$ may introduce function constants but is not recursive. Therefore, every bottom–up evaluation of $P \cup V^- \cup D$ progresses in two stages. The first stage produces instances of the views in $V^-$. The second stage is a standard Datalog evaluation of $P$ on the resulting dataset. Because every Datalog program has a unique minimal model, this proves the claim. ∎

Given instances for all base relations of a logic program, the program resulting from inversion might produce tuples containing function constants in its result. Because instances of base relations do not contain function constants, no Datalog program produces output data that contain function constants. This motivates the definition of a filter that gets rid of all such data. If $D$ is a database containing instances of the base relations of a logic program $P$, then let $P(D)\!\downarrow$ be the set of all tuples in $P(D)$ that do not contain any function constants.

Let $P\!\downarrow$ be a logic program that, given a database $D$, computes $P(D)\!\downarrow$. A program of this sort can be formed by writing rules that define a new query relation in terms of the original query relation by succeeding only on those tuples that do not contain functional terms.

The following theorem asserts that adding inverses of conjunctively defined source relations yields a logic program that uses the views in the best possible way. After discarding all tuples containing function constants, the result is contained in $P$. Moreover, the result of every retrievable query is already contained in this program as well.

**Theorem:** For every Datalog program $P$ and every set of conjunctively defined views $V$, $(P \cup V^-)\!\downarrow$ is maximally contained in $P$. Moreover, $P \cup V^-$ can be constructed in time that is polynomial in the sizes of $P$ and $V$.

**Proof:** First, we prove that $(P \cup V^-)\!\downarrow$ is contained in $P$. Let $B_1, \ldots, B_n$ be instances of the base relations in $P$. $B_1, \ldots, B_n$ determine instances of the views in $V$, which, in turn, are the input relations for $(P \cup V^-)\!\downarrow$. Assume that $(P \cup V^-)\!\downarrow$ produces a proposition $p$ that does not contain any functional terms. Consider the derivation tree of $p$ in $(P \cup V^-)\!\downarrow$. All of the leaves are view literals because view relations are the only input relations of $(P \cup V^-)\!\downarrow$. Removing the leaves from this tree produces a new tree with the input relations from $P$ as the new leaves. Because the instances of the views are derived from $B_1, \ldots, B_n$, there are constants in $B_1, \ldots, B_n$ such that consistently replacing the functional terms with these constants yields a derivation tree of $p$ in $P$. Therefore, $(P \cup V^-)\!\downarrow$ is contained in $P$.

Let $P'$ be an arbitrary retrievable query plan contained in $P$. We have to prove that $P'$ is also contained in $(P \cup V^-)\!\downarrow$. Let $C'$ be an arbitrary conjunctive query generated by $P'$. If we can prove that $C'^{\infty}$ is contained in $(P \cup V^-)\!\downarrow$, then $P'$ is contained in $P$. Let $D$ be the canonical database of $C'$. Because $C'^{\infty}$ is contained in $P$, $C'^{\infty}(D)$ is contained in the output of $P$ applied to $D$. Let $C''$ be the conjunctive query generated by $P$ that produces $C'^{\infty}(D)$. Because all relations of query $C'$ are also in $C'^{\infty}$ and all relations in $C'^{\infty}$ appear in some view definition, $C$ is also generated by $P$. Because $C'$ is contained in $C$, there is a containment mapping $h$ from $C$ to $C'^{\infty}$. Every variable $Z$ in $C'$ that does not appear in $C'$ is existentially quantified in some view definition in $C'$ with $v'(X_1, \ldots, X_n)$ as head. Let $k$ be the mapping that maps every such variable to the corresponding term $f(X_1, \ldots, X_n)$ used in the inverse of the view. Because $P$ can derive $C$, $P$ can also derive the more specialized conjunctive query $k(h(C))$. Using rules in $V^-$, the derivation of $k(h(C))$ in $P$ can be extended to a derivation of a conjunctive query $C''$ that contains only view literals. The identity mapping is a containment mapping from $C''$ to $C'$. This proves that $P'$ is contained in $(P \cup V^-)\!\downarrow$.

$(P \cup V^-)\!\downarrow$ can be constructed in time polynomial in the size of $P$ and $V$ because every subgoal in a view definition in $V$ corresponds to exactly one inverse rule in $V^-$. ∎

One interesting upshot of this observation is that it is possible to transform every Functional Datalog program produced by the Inverse Method into an equivalent Datalog program without functions. The transformation, called *flattening*, proceeds in a bottom–up fashion. Functional terms like $f(X_1, \ldots, X_n)$ are eliminated by replacing them with the variables $X_1, \ldots, X_n$ and by changing the relation within which they occur to a new relation that indicates that those variables belong to the functional term $f(X_1, \ldots, X_n)$.

As an example of this transformation, consider the rule shown below.

```
flight(C,g(C,F,T),T)  :-  v3(C,F,T)
```

In this case, the result is as follows. The subscript here represents the fact that the first argument of `flight` corresponds to the first argument of `flight_1_g_2_3_4_5`, the second argument of `flight` corresponds to the term consisting of the function constant g applied to the second, third, and fourth arguments of `flight_1_g_2_3_4_5`, and the third argument of `flight` corresponds to the fifth argument of `flight_1_g_2_3_4_5`. Note that the arity of each function constant in a suffix (e.g., the g in `flight_1_g_2_3_4_5`) determines how many of the following suffixes are associated with that function constant (in this case, 3).

```
flight_1_g_2_3_4_5(C,F,C,T,T) :- v3(C,F,T)
```

In the following, let us denote the Datalog program resulting from this transformation by $P^*$. Clearly, there is a one-to-one correspondence between bottom–up evaluations in $P$ and in $P^*$. Because we keep track of function constants in $P^*$, we know that the resulting instance of the query relation is exactly the subset of the result of $P$ that does not contain function constants.

Before we leave our discussion of conjunctively defined source relations, let us look at a slightly more interesting example. In this case, although the source relations are conjunctive, the query relation is disjunctive and recursive.

Consider the Datalog program shown below. It defines the transitive closure of the binary relation edge.

```
query(X,Y) :- edge(X,Y)
query(X,Y) :- edge(X,W) & query(W,Y)
```

Assume there is only one view available, which relates the endpoints of paths of length two.

```
v(X,Y) :- edge(X,W) & edge(W,Y)
```

Just using this view, there is no way to determine the transitive closure of the edge relation. The best we can hope to achieve is to compute the endpoints of paths of even length.

In this case, the Inverse Method produces the following Functional Datalog program.

```
query(X,Y) :- edge(X,Y)
query(X,Y) :- edge(X,W) & query(W,Y)
edge(X,f(X,Y)) :- v(X,Y)
edge(f(X,Y),Y) :- v(X,Y)
```

This program indeed yields all endpoints of paths of even length in its result.

## 3.5    DISJUNCTIVE SOURCE DEFINITIONS

In this section, we examine the query-folding problem in the presence of disjunctively defined source relations. As we shall see, this case is more complex than the case of conjunctively defined source relations. Nevertheless, there is an extension to the Inverse Method that generates retrievable query plans guaranteed to extract all available information from disjunctively defined sources.

We start by showing that Basic Datalog is not sufficiently expressive in this case. We also show that Disjunctive Datalog enhanced with inequalities is okay.

Assume an information source stores flight information. More precisely, the source stores nonstop flights by United Airlines (ua) and Southwest Airlines (sw), and flights out of San

Francisco International Airport (sfo) with one stopover. The information stored by this source can be described as being a view over a database with a base relation flight that stores all nonstop flights.

```
v(ua,From,To)  :-  flight(ua,From,To)
v(sw,From,To)  :-  flight(sw,From,To)
v(A,sfo,To)  :-  flight(A,sfo,Stop) & flight(A,Stop,To)
```

Now suppose that a user is interested in all cities with nonstop flights to Los Angeles. He could express this query as shown below.

```
query(From)  :-  flight(Airline,From,lax)
```

If v(ua,jfk,lax) is a datum in the source relation, then there is clearly a nonstop flight from New York (jfk) to Los Angeles (lax). On the other hand, if the tuple v(ua,sfo,lax) is in the relation, then there does not necessarily exist a nonstop flight from San Francisco to Los Angeles. Indeed, this tuple could be included because there is a flight with one stopover from San Francisco to Los Angeles. The task of query planning in information integration systems is to find a query plan, i.e., a query that requires only views and that extracts as much information as possible from the available sources. All flights to Los Angeles in the relation with the exception of flights departing from San Francisco International Airport are nonstop flights. Therefore, the following query plan produces all of the guaranteed answers to this query.

```
query(From)  :-  v(Airline,From,lax) & From!=sfo
```

Note that, without the use of inequality, it would not be possible to guarantee that all cities returned by the query plan have nonstop flight to Los Angeles without enumerating all possible cities. Unfortunately, just adding inequality is not enough for some queries and views, as shown by the following example.

Assume that there are two source relations available, which are described by the following view definitions. View v1 stores vertices that are colored red, green, or blue. View v2 stores pairs of vertices that are connected by an edge.

```
v1(X)  :-  color(X,red)
v1(X)  :-  color(X,green)
v1(X)  :-  color(X,blue)

v2(X,Y)  :-  edge(X,Y)
```

Now, suppose that the user wants to know whether there is a pair of vertices of the same color that are connected by an edge. The following query expresses this request.

```
query(yes)  :- edge(X,Y) & color(X,Z) & color(Y,Z)
```

Consider the graphs $G_1$, $G_2$, and $G_3$ shown below. All of these graphs are non-three-colorable, i.e., for every possible coloring of the vertices with at most three colors, there will be one edge that connects vertices with the same color.

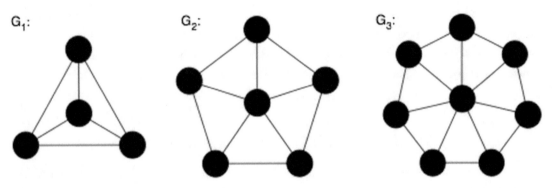

FIGURE 3.1: Graphs.

Therefore, every graph that contains $G_1$, $G_2$, or $G_3$ as a subgraph contains an edge that connects two vertices with the same color if the vertices in $G_1$, $G_2$, and $G_3$ are colored with at most three colors. These conditions can be easily checked by a Datalog query plan. For example, the plan shown below outputs yes exactly when the input graph contains $G_1$ as a subgraph and when the vertices are colored with at most three colors.

```
query(yes)  :-     v1(X1) & v1(X2) & v1(X3) & v1(Y) &
                   v2(X1,X2) & v2(X2,X3) & v2(X3,X1) &
                   v2(X1,Y) & v2(X2,Y) & v2(X3,Y)
```

It follows that this query plan is contained in our query. More generally, every query plan that checks that the input graph contains a non-three-colorable subgraph and that all the vertices in the subgraph are colored by at most three colors is contained in this query plan. Also, every query plan that is contained in the query outputs yes only if the graphs described by v1 and v2 are non-three-colorable.

It is well known that deciding whether a graph is three-colorable is NP-complete. Because the problem of evaluating a Datalog program has polynomial data complexity, this shows that there is no Datalog query plan that contains all the query plans that are contained in our query. Intuitively, the reason is that, for every Datalog query plan $P$ that is contained in $Q$, an additional conjunctive query that tests for one more non-three-colorable graph can be added to create a query plan that is still contained in $Q$ but is not contained in $P$.

On the other hand, it *is* possible to write a retrievable query plan in disjunctive Datalog. The following plan suffices.

```
query(yes)  :- v2(X,Y) & color(X,Z) & color(Y,Z)
color(X,red) | color(X,green) | color(X,blue)  :- v1(X)
```

In this section, we present a construction that produces maximally contained query plans in the presence of disjunctively defined sources. The central part of our construction is the concept of disjunctive inverse rules. Before we can proceed to this definition, we have to define some technical concepts. Let $\{r_1, \ldots, r_n\}$ be a positive view definition of the form shown below.

$$v(\bar{X}_1)  :-  u_{11}(\bar{X}_{11})  \&  \ldots  \&  u_{1m_1}(\bar{X}_{1m_1}) \tag{$r_1$}$$

$$\ldots$$

$$v(\bar{X}_n)  :-  u_{n1}(\bar{X}_{n1})  \&  \ldots  \&  u_{nm_n}(\bar{X}_{nm_n}) \tag{$r_n$}$$

We can assume without loss of generality that the vectors $\bar{X}_1, \ldots, \bar{X}_n$ are mutually disjoint. Given an instance $d$ of relation $v$, we have to determine which of the rules $r_1, \ldots, r_n$ might have generated $d$. If there is a datum $d$ such that $d$ can be generated by every one of the rules $r_{i_1}, \ldots, r_{i_k}$, then these rules are called *truly disjunctive*. Viewed another way, rules $r_{i_1}, \ldots, r_{i_k}$ are called truly disjunctive if and only if there is a substitution $\sigma$ such that $\bar{X}_{i_1}\sigma=\ldots=\bar{X}_{i_k}\sigma$. In this case, $\bar{X}_{i_1}\sigma$ is called a *witness* of $r_{i_1}, \ldots, r_{i_k}$ being truly disjunctive. A witness $\bar{X}$ is *most general* if and only if there is no other witness that can be converted to $\bar{X}$ by substituting expressions for any variables in that witness.

Let the arity of $v$ be $m$, and let $c_1, \ldots, c_m$ be new constants. A conjunction of inequalities $\phi$ involving only $c_1, \ldots, c_m$ and the constants in $\bar{X}_1, \ldots, \bar{X}_n$ is called an *attribute constraint*. A conjunctive rule $r_i$ satisfies an attribute constraint $\phi$ if and only if all inequalities in $\phi$ hold after replacing each $c_j$ in $\phi$ by the corresponding $\bar{X}_i[j]$. If rules $r_{i_1}, \ldots, r_{i_k}$ are truly disjunctive with most general witness $Z$ and there is an attribute constraint $\phi$ satisfied by $\bar{X}_1, \ldots, \bar{X}_n$ but not satisfied by any $X_j$ with $j \in \{1, \ldots, n\}$-$\{i_1, \ldots, i_k\}$ and $X_j$ unifiable with $Z$, then $\phi$ is called an *exclusion condition* for $r_{i_1}, \ldots, r_{i_k}$.

We are now able to define the central concept of disjunctive inverse rules. Intuitively, inverse rules describe all the databases that are consistent with the view definitions given a specific view instance.

Consider the rules $\{r_1, \ldots, r_n\}$ in the view definition shown above. Assume that the head variables in each rule are disjoint. Let $\bar{X}_1, \ldots, \bar{X}_s$ be those variables that appear in the bodies but not in the heads, and let $f_1, \ldots, f_s$ be new function constants. Then, for every set of truly disjunctive rules $r_{i_1}, \ldots, r_{i_k}$ with most general witness $\bar{Z}$ and most general exclusion condition $\phi$, we have the following disjunctive inverse rules.

$$u_{i_1 \delta_1}(\bar{X}'_{i_1 \delta_1}) \quad | \quad \ldots \quad | \quad u_{i_k \delta_k}(\bar{X}'_{i_k \delta_k}) \quad :- \quad v(\bar{Z}) \quad \& \quad \phi'$$

Here, we have a separate rule for every combination of values for $\delta_l$, where $\delta_l \in \{1, \ldots, m_l\}$ for $l \in \{1, \ldots, k\}$ and where the following conditions hold for all $\beta \in \{i_1, \ldots, i_k\}$, for all $\gamma \in \{1, \ldots, m\}$, and for all $j$ from 1 through the arity of $u_{\beta \gamma}$.

| | | |
|---|---|---|
| $X'_{\beta\gamma}[j]=Z[j']$ | if | $X_{\beta\gamma}[j]=X_{\beta}[j']$ for some $j' \in \{1, \ldots, m\}$ |
| $X'_{\beta\gamma}[j]=X_{\beta\gamma}[j]$ | if | $X_{\beta\gamma}[j]$ is a constant |
| $X'_{\beta\gamma}[j]=f_i(Z)$ | If | $X_{\beta\gamma}[j]=X_i$ for some $i \in \{1, \ldots, s\}$ |

Condition $\phi'$ is generated from $\phi$ by replacing each constant in $\phi$ by the corresponding variable or constant.

As a simple example of this method, consider an application in which there is a single disjunctively defined view v with the definition shown below.

```
v(X)  :-  p(X)
v(c)  :-  q(c)
```

Our job in this case is to find a retrievable query plan that computes values for the relation p.

```
query(X)  :-  p(X)
```

Using the Inverse Method to invert the view definitions, we end up with the following inverse rules.

```
p(X)  :-  v(X)  &  X!=c
q(c)  :-  v(c)
p(c)  |  q(c)  :-  v(c)
```

Combining these with the view definition and simplifying, we end up with the following retrievable query plan. This plan is not equivalent to the original query, since it does not say anything about the entity c. However, it is maximally contained in the original query; no plan is better since we have no information about the relation q.

```
query(X)  :- v(X)  & X!=c
```

As a more elaborate example of the method, consider the source definitions in the first example in this section. The disjunctive inverse rules in this case are the following.

```
flight(ua,F1,T1)  :- v(ua,F1,T1)  & F1 != sfo
flight(ua,F2,T2)  :- v(sw,F2,T2)  & F2 != sfo
flight(A3,sfo,f(A3,sfo,T3))  :- v(A3,sfo,T3)  & A3 != ua & A3 != sw
flight(A3,f(A3,sfo,T3),T3)  :- v(A3,sfo,T3)  & A3 != ua & A3 != sw
flight(ua,sfo,T1)  | flight(ua,sfo,f(ua,sfo,T1))  :- v(ua,sfo,T1)
flight(ua,sfo,T1)  | flight(ua,f(ua,sfo,T1),T1)  :- v(ua,sfo,T1)
flight(ua,sfo,T1)  | flight(sw,sfo,f(sw,sfo,T2))  :- v(sw,sfo,T2)
flight(ua,sfo,T1)  | flight(sw,f(sw,sfo,T2),T2)  :- v(sw,sfo,T2)
```

Once again, combining these rules with the original query and simplifying, we end up with the maximally contained retrievable query plan equivalent to the one described earlier.

```
query(From)  :- v(ua,From,lax)  & From != sfo
query(From)  :- v(sw,From,lax)  & From != sfo
```

In the following, we consider the query plan consisting of the rules of the Datalog query $Q$ together with the disjunctive inverse rules $V^-$. Disjunctive inverse rules can contain function constants. Therefore, the output of our query plan $Q \cup V^-$ on a specific database can contain tuples with function constants. Fortunately, we can transform $Q \cup V^-$ to $(Q \cup V^-)\!\downarrow$ using a transformation akin to the one described in the preceding section.

The following theorem guarantees that the query plan $(Q \cup V^-)\!\downarrow$ is guaranteed to be maximally contained in $Q$.

**Theorem:** For every Datalog query $Q$ and every set of positive view definitions $V$, the disjunctive Datalog query plan $(Q \cup V^-)\!\downarrow$ is maximally contained in $Q$.

**Proof:** Let $I$ be a view instance. Because the $Q$ part of query plan $Q \cup V^-$ does not contain any source relations and because all the relations in the bodies of $V^-$ are source relations, every bottom-up evaluation of $Q \cup V^-$ must necessarily evaluate $V^-$ before evaluating $Q$. Therefore, $(Q \cup V^-)(I)$ is the intersection of $Q(M)$ for all $M$ that satisfy $V^-(I)$.

Let $A$ be the intersection of $Q(M)$ for all $M$ that satisfy $V^-(I)$ after filtering out all functional terms and let $B$ be the intersection of $Q(D)$ for all $D$ such that $I \subseteq V(D)$. Since the disjunctive

Datalog queries are monotone, it suffices to show that $A=B$. Let $M$ be a model of $V^-$ and $I$. By the construction of $V^-$, we know that $I \subseteq V(M)$. Therefore, $B \subseteq A$. Because $B$ does not contain function constants, it follows that $B \subseteq A\!\!\downarrow$.

Let $D$ be a database with $I \subseteq V(D)$. Consider all the models of $V^-$ and $V(D)$. One of these models must coincide with $D$ with the only difference being that some of the functional terms are replaced by constants in $D$. Let $M$ be this model. Because Datalog queries are monotone, when constants in the input database are made equal, it follows that $Q(M) \subseteq Q(D)$. Therefore, $A\!\!\downarrow \subseteq B$. ∎

**Theorem:** For every Datalog query $Q$ and every set of positive view definitions $V$, the disjunctive Datalog query plan $(Q \cup V^-)\!\!\downarrow$ can be evaluated in co-NP time.

**Proof:** Let $d$ be a datum that is not in $(Q \cup V^-)\!\!\downarrow(I)$ for some instance $I$. Then there is some model $M$ of $V^-$ and $I$ such that $d$ is not in $Q(M)$. If $M$ contains $n$ items and the longest conjunct in $V$ has $m$ literals, then there is a submodel $M'$ of $M$ with at most $nm$ members that is still a model of $V^-$. Because of the monotonicity of $Q$, $d$ is also not in $M'$. Moreover, checking that $M'$ is also a model of $V^-$ and that $d$ is not in $Q(M)$ can be done in polynomial time. ∎

## 3.6    COMPARISON TO OTHER METHODS

In this section, we compare the Inverse Method to the Bucket Algorithm and the Unijoin Algorithm. Unlike the Inverse Method, these algorithms can handle only conjunctive queries and conjunctive source relations. We illustrate our comparison on a simple conjunctive problem, i.e., the example introduced in Section 3.3.

### Bucket Algorithm

The Bucket Algorithm applies to conjunctively defined views only. For each subgoal $p_i$ in the user query, a bucket $B_i$ is created. If $v_j$ is the head of a rule containing a subgoal that unifies with $p_i$ with unifier $\sigma$, then $v_j\sigma$ is inserted into $B_i$. Given a conjunctive query $c$, the bucket algorithm constructs retrievable conjunctive query plans with the same head as $c$ and all possible conjunctions of items in the buckets as bodies. For each of these retrievable query plans, the algorithm checks whether it can add a constraint to the body such that the expansion of the resulting query is contained in $c$. All retrievable query plans that pass this test are then evaluated to construct an answer to the user's query.

Once again, consider the example introduced in Section 3.3. In this case, the Bucket Algorithm creates three buckets $B_1$, $B_2$, and $B_3$ for the three subgoals of `flight(C,tus,sfo)`, `flight(C,tus,Z)`, and `flight(C,Z,sfo)`, respectively. The buckets are filled as follows.

| $B_1$ | $B_2$ | $B_3$ |
|---|---|---|
| v1(tus,sfo) | v1(tus,Z) | v1(Z,sfo) |
| v2(tus,sfo) | v2(tus,Z) | v2(Z,sfo) |
| v3(C,tus,T1) | v3(C,tus,T2) | v3(C,Z,T3) |
| v3(C,F1,sfo) | v3(C,F2,Z) | v3(C,F3,sfo) |

The retrievable queries for the first conjunctive query are shown below. For each of these, the Bucket Method checks whether, after adding some constraints, its expansion is contained in the first conjunctive query.

$$q(C) \ :- \ v1(tus,sfo) \tag{1}$$
$$q(C) \ :- \ v2(tus,sfo) \tag{2}$$
$$q(C) \ :- \ v3(C,tus,T1)$$
$$q(C) \ :- \ v3(C,F1,sfo)$$

The following sixteen retrievable queries stem from the second conjunctive query. Each is checked to see whether, after adding some constraints, its expansion is contained in the second conjunctive query.

$$q(C) \ :- \ v1(tus,Z) \ \& \ v1(Z,sfo) \tag{3}$$
$$q(C) \ :- \ v1(tus,Z) \ \& \ v2(Z,sfo)$$
$$q(C) \ :- \ v1(tus,Z) \ \& \ v3(C,Z,T3) \tag{4}$$
$$q(C) \ :- \ v1(tus,Z) \ \& \ v3(C,F3,sfo) \tag{5}$$
$$q(C) \ :- \ v2(tus,Z) \ \& \ v1(Z,sfo)$$
$$q(C) \ :- \ v2(tus,Z) \ \& \ v2(Z,sfo) \tag{6}$$
$$q(C) \ :- \ v2(tus,Z) \ \& \ v3(C,Z,T3) \tag{7}$$
$$q(C) \ :- \ v2(tus,Z) \ \& \ v3(C,F3,sfo) \tag{8}$$
$$q(C) \ :- \ v3(C,tus,T2) \ \& \ v1(Z,sfo)$$
$$q(C) \ :- \ v3(C,tus,T2) \ \& \ v2(Z,sfo)$$
$$q(C) \ :- \ v3(C,tus,T2) \ \& \ v3(C,Z,T3) \tag{9}$$
$$q(C) \ :- \ v3(C,tus,T2) \ \& \ v3(C,F3,sfo) \tag{10}$$
$$q(C) \ :- \ v3(C,F2,Z) \ \& \ v1(Z,sfo) \tag{11}$$
$$q(C) \ :- \ v3(C,F2,Z) \ \& \ v2(Z,sfo) \tag{12}$$

$$q(C) \quad :- \quad v3(C,F2,Z) \quad \& \quad v3(C,Z,T3) \tag{13}$$
$$q(C) \quad :- \quad v3(C,F2,Z) \quad \& \quad v3(C,F3,sfo) \tag{14}$$

For each numbered retrievable query, a constraint can be added to its body such that it passes the containment test. For example, the constraint that needs to be added to query (1) is C=wn, and the constraint that needs to be added to query (10) is T=sfo. The resulting query set is shown below.

```
q(C)  :-  v1(tus,sfo)
q(C)  :-  v2(tus,sfo)
q(C)  :-  v3(C,tus,T1)
q(C)  :-  v3(C,F1,sfo)
```

As the example shows, the Bucket Method performs a lot of containment tests. This is quite expensive, especially because testing containment of conjunctive queries is NP-complete.

## Unijoin Algorithm

The first step of the Unijoin Algorithm is the same as the first step of the Inverse Method, i.e., the generation of inverse rules. This is not surprising, as the Inverse Method was inspired in part by the Unijoin Algorithm. However, after this first step, there is a difference. Where the Inverse Method transforms the original query together with the inverse rules into a retrievable Datalog program, the Unijoin Algorithm constructs a set of retrievable conjunctive queries.

For each subgoal $p_i$ in the user query, a label $L_i$ is created. If $r \;:-\; v$ is one of the inverse rules and $r$ and $p_i$ are unifiable, then the pair $<\sigma{\downarrow}p_i, v\sigma>$ is inserted into $L_i$ provided that $\sigma{\downarrow}q$ does not contain any functional terms. Here, $\sigma$ is a most general unifier of $r$ and $p_i$ and $\sigma{\downarrow}p_i$ and $\sigma{\downarrow}q$ are restrictions of $\sigma$ to the variables in $p_i$ and $q$. The *unification–join* of two labels $L_1$ and $L_2$ is defined as follows. If $L_1$ contains a pair $<\sigma_1,t_1>$ and $L_2$ contains a pair $<\sigma_2,t_2>$, then the unification–join of the two contains the pair $<\sigma_1\sigma\cup\sigma_2\sigma, t_1\&t_2>$, where $\sigma$ is a most general substitution such that $\sigma_1\sigma{\downarrow}\sigma_2=\sigma_2\sigma{\downarrow}\sigma_1$ provided there is such a substitution $\sigma$ and provided that $\sigma_1\sigma{\downarrow}q$, $\sigma_2\sigma{\downarrow}q$, and $(t_1 \& t_2)\sigma$ do not contain any functional terms. If $<\sigma,v_1\&\ldots\&v_n>$ is in the unification–join of all labels corresponding to the subgoals in one of the conjunctive user queries and this user query has head $h$, then the retrievable query with head $h\sigma$ and body $v_1\&\ldots\&v_n$ is part of the result.

Applied to the query above, the Unijoin Algorithm generates three labels, namely, $L_1$ and $L_2$ and $L_3$, which correspond to the subgoals `flight(C,tus,sfo)`, `flight(C,tus,Z)`, and `flight(C,Z,sfo)`. The contents of these labels are shown below.

```
<{C←wn}, v1(tus,sfo)>
<{C←ua}, v2(tus,sfo)>
```

```
<{C←wn}, v1(tus,Z)>
<{C←ua}, v2(tus,Z)>
<{Z←g(C,tus,T)}, v3(C,tus,T)>

<{C←wn}, v1(Z,sfo)>
<{C←ua}, v2(Z,sfo)>
<{Z←g(C,F,sfo)}, v3(C,F,sfo)>
```

The unification–join of $L_2$ and $L_3$ is the following.

```
<{C←wn}, v1(tus,Z) & v1(Z,sfo)>
<{C←ua}, v1(tus,Z) & v1(Z,sfo)>
<{Z←g(C,tus,sfo)}, v3(C,tus,sfo)>
```

The labels corresponding to the conjunctive queries are $L_1$ and the join of $L_2$ and $L_3$, respectively. The retrievable conjunctive queries that can be constructed from these are the following.

```
query(wn) :- v1(tus,sfo)
query(ua) :- v2(tus,sfo)
query(wn) :- v1(tus,Z) & v1(Z,sfo)
query(wn) :- v2(tus,Z) & v2(Z,sfo)
query(C) :- v3(C,tus,sfo)
```

The Unijoin Algorithm does not require containment tests. On the other hand, it can generate an exponential number of conjunctive queries in cases where the Inverse Method generates a small Datalog program.

As an example, assume that there are $k$ views defined as shown below.

$$v_1(X, Y) \quad :- \quad u(X, Y), \quad u_1(X, Y)$$
$$\cdots$$
$$v_k(X, Y) \quad :- \quad u(X, Y), \quad u_k(X, Y)$$

Now consider a query of the following form.

$$\mathtt{query}(X_0, X_n) \quad :- \quad u(X_0, X_1) \quad \& \quad \cdots \quad \& \quad u(X_{n-1}, X_n)$$

In this case, the Unijoin Algorithm produces the following plan. Every view appears in every position in all possible combinations.

```
query (X₀, Xₙ)   :-   v₁ (X₀, X₁)   &  ...  &  v₁ (Xₙ₋₁, Xₙ)
query (X₀, Xₙ)   :-   v₁ (X₀, X₁)   &  ...  &  v₂ (Xₙ₋₁, Xₙ)
                                    . . .
query (X₀, Xₙ)   :-   vₖ (X₀, X₁)   &  ...  &  vₖ (Xₙ₋₁, Xₙ)
```

Given the same inputs, the Inverse Method constructs the following Datalog program.

```
query (X₀, Xₙ)   :-   u (X₀, X₁)   &  ...  &  u (Xₙ₋₁, Xₙ)
u (X, Y)   :-   v₂ (X, Y)
                       . . .
u (X, Y)   :-   vₙ (X, Y)
```

Evaluating the unijoin plan requires $(n-1)k^n$ joins and $k^n-1$ unions. By contrast, evaluating the inverse plan requires just $k-1$ unions and $n-1$ joins, which is significantly better.

## 3.7   DECIDABILITY

In this section, we show the limits of answering queries using views. In the case of purely conjunctive queries and views, the problem is in NP. In the case of recursive queries, even with conjunctive views, the question of the existence of an *equivalent* retrievable query plan is undecidable.

**Theorem:** Given an open Datalog program P with base relations $b_1, \ldots, b_m$ and conjunctive source relations defined in terms of $b_1, \ldots, b_m$, it is undecidable whether there is a retrievable Datalog program that is equivalent to $P$.

**Proof:** Let $P_1$ and $P_2$ be arbitrary Datalog programs. We show that a decision procedure for the problem above would allow us to decide whether $P_1$ is contained in $P_2$. Because the containment problem for Datalog programs is undecidable, this proves the claim. Without loss of generality, we can assume that there are no views with the same name in $P_1$ and $P_2$ and that the answer relations in $P_1$ and $P_2$ have arity $m$ and are named query1 and query2, respectively. Let $P$ be the Datalog program consisting of all of the rules in $P_1$ and $P_2$ together with the following rules, where e is a new 0-ary relation constant.

```
query (X₁, ..., Xₘ)   :-   query1 (X₁, ..., Xₘ)   &  e ()
query (X₁, ..., Xₘ)   :-   query2 (X₁, ..., Xₘ)
```

For every base relation $e_i (X_1, \ldots, X_{ki})$ in $P_1$ and $P_2$, but not for e, we define the following view.

```
vᵢ (X₁, ..., Xₖᵢ)   :-   eᵢ (X₁, ..., Xₖᵢ)
```

We show that $P_1$ is contained in $P_2$ if and only if there is a retrievable Datalog program $P'$ equivalent to $P$.

Assume that $P_1$ is contained in $P_2$. Then $P$ is equivalent to the program $P'$ consisting of all of the rules of $P_2$ with each $b_i$ replaced by the corresponding $v_i$. together with the additional rule shown below.

$$\texttt{query}\,(X_1,\ldots,X_m) \quad \texttt{:-} \quad \texttt{query2}\,(X_1,\ldots,X_m)$$

Now, assume that there is a retrievable Datalog program $P'$ equivalent to $P$. Then for any instantiation of the base relations, $P$ and $P'$ yield the same result, especially for instantiations where e is the empty relation and where e contains the empty tuple. If e is the empty relation, then $P$ produces exactly the tuples produced by $P_2$; and, therefore, $P'$ does likewise. If e contains the empty tuple, then $P$ produces the union of the tuples produced by $P_1$ and $P_2$; and, hence, $P'$ produces this union. Because $P'$ does not contain e(), $P'$ will produce the same set of tuples regardless of the instantiation of e. It follows that $P_2$ is equivalent to the union of $P_1$ and $P_2$. Therefore, $P_1$ is contained in $P_2$. ◆

C H A P T E R   4

# Query Planning

## 4.1   INTRODUCTION

A retrievable query plan defines an overall query relation in terms of available source relations. However, it does not specify exactly how that query is to be evaluated. In this chapter, we discuss the various decisions a broker must make, and we describe some approaches to making these decisions.

The next section discusses how to optimize a retrievable query plan to enhance the efficiency of evaluation. Section 4.3 describes the grouping of queries and the selection of sources for evaluating these groups. Section 4.4 deals with the choice of method for executing query groups.

## 4.2   OPTIMIZATION

In the presence of database constraints, it is sometimes possible to enhance the efficiency of retrievable query plans without sacrificing answers. In this section, we look at some techniques for using constraints to optimize retrievable query plans, in particular, pruning empty conjunctions, deleting redundant conjuncts from conjunctions, and deleting unnecessary disjuncts from disjunctions.

As an example of the need for pruning empty conjunctions, consider a data integration problem with a single base relation `flight` that relates every nonstop flight, the airline offering that flight, the origin, the destination, a Boolean indicating whether the flight has a movie, and the type of aircraft used for that flight.

Assume that there are four source relations. The first `ua_movie` lists United Airlines flights that offer movies. The second `aa_movie` lists American Airlines flights that offer movies. The third `ua_aircraft` lists United Airlines flights together with the aircraft used on those flights. The fourth `aa_aircraft` lists analogous information for American Airlines flights. The definitions for these source relations in terms of flight are shown below.

```
ua_movie(N)  :- flight(N,ua,X,Y,yes,A)
aa_movie(N)  :- flight(N,aa,X,Y,yes,A)
ua_aircraft(N,A)  :- flight(N,ua,X,Y,M,A)
aa_aircraft(N,A)  :- flight(N,aa,X,Y,M,A)
```

---

*An interactive version of this book is available that includes implementations of algorithms used in demonstrations and exercises. See page xi for more information.*

Now, suppose that the user is interested in finding all flights on Boeing 737s that offer movies. The query is shown below.

```
query(C,N)  :-  flight(N,C,X,Y,yes,737)
```

The Inverse Method in this case produces the retrievable query plan shown below.

```
query(C,N)  :-  ua_movie(N)  &  ua_aircraft(N,737)
query(C,N)  :-  ua_movie(N)  &  aa_aircraft(N,737)
query(C,N)  :-  aa_movie(N)  &  ua_aircraft(N,737)
query(C,N)  :-  aa_movie(N)  &  aa_aircraft(N,737)
```

Without further constraints, this plan is as small as possible. None of the conjunctions are necessarily empty, and so the entire plan must be evaluated. However, in reality, it is unlikely that the United Airlines relation would contain information about the American Airlines flights and vice versa. Pruning away such cases is easy if constraints among the relations are available to the data integrator. One way to write the constraints is shown below.

```
illegal  :-  ua_movie(N)  &  aa_movie(N)
illegal  :-  ua_movie(N)  &  aa_aircraft(N,A)
illegal  :-  ua_aircraft(N,A)  &  aa_movie(N)
illegal  :-  ua_aircraft(N,A)  &  aa_aircraft(N,B)
```

Detecting and eliminating empty conjunctions with constraints expressed in this form is easy. We simply check each conjunction to see whether it is contained in the body of a constraint, i.e., we see whether there is some mapping of variables from the body of the constraint to constants and variables in the query that makes it a subset of the query conditions. If so, then the conjunction is empty and can be eliminated.

In our example here, notice that the second constraint eliminates the second conjunction in the query, and the third constraint eliminates the third conjunct. As a result, both conjunctions can be dropped, and we end up with the leaner, more efficient query plan shown below.

```
query(C,N)  :-  ua_movie(N)  &  ua_aircraft(N,737)
query(C,N)  :-  aa_movie(N)  &  aa_aircraft(N,737)
```

One problem with this method is that it assumes that constraints are written in terms of source relations. In practice, the constraints are usually written in terms of base relations or some mix of base and source relations.

For example, in this case, all of the constraints could be captured via a key constraint on the flight relation, expressed as shown below.

```
illegal :-
    flight(N,C1,X1,Y1,M1,A1) &
    flight(N,C2,X2,Y2,M2,A2) &
    C1!=C2
```

Of course, this constraint does not apply to the query plan directly. However, it is possible to generate specific constraints from this more general constraint by taking the deductive closure of all constraints and all view definitions. Constraints like the ones above would then be found amid these conclusions.

The downside of this approach is that the deductive closure can sometimes be very large. An alternative that deals with this size at the cost of slightly more expensive consistency checking is to leave constraints as they are and use automated reasoning methods to deduce a contradiction between the assumption of an answer to a conjunction and the constraint set.

## 4.3  SOURCING

A *retrievable query plan* defines an overall query relation in terms of available source relations. However, it does not specify which queries are to be sent to the available sources. In the case of replication of relations among multiple sources, there is a choice of which source to ask in retrieving data from a given relation. In the case of a single source plan, there is a choice of whether to ask one single query or lots of little queries.

An *executable query plan* is a reformulation of a retrievable query plan in which all of these choices are spelled out. There are two components. First, there are definitions of the *source queries*, each of which is defined entirely in terms of the relations available at a single source. Second, there is the *local query plan*, which defines the overall query relation in terms of these source queries. *Sourcing* is the process of converting a retrievable query plan into an equivalent executable query plan.

As an example of this process, consider the following retrievable query plan.

```
query(X,Z) :- p(X) & q(X,Y) & s(Y,Z)
query(X,Z) :- r(1,X,Z,W)
```

Assume that there are two sources of data: Xanadu and Yankee. Xanadu hosts relations p, q, and r, while Yankee hosts p and s. The following is an executable query plan for this case.

```
query(X,Y) :- xanadu_1(X,Y) & yankee_1(Y,Z)
query(X,Y) :- xanadu_2(X,Y)
```

```
xanadu_1(X,Y) :- p(X) & q(X,Y)
xanadu_2(X,Z) :- r(1,X,Z,W)
yankee_1(Y,Z) :- s(Y,Z)
```

Choosing among alternative executable query plans requires that we settle on a measure to be optimized. Overall computation time is an obvious candidate. Communication cost is another. In what follows, we assume that the computation cost is similar no matter which computer executes the computation. In the Internet setting, communication cost is often more significant. There is cost in setting up connections to sources and cost in moving large datasets between computers. In what follows, we concentrate primarily on communication cost.

We begin our discussion of sourcing with the case of a retrievable query plan in which all relations are hosted at a single source.

As mentioned above, one approach to sourcing in this case is to send the entire retrievable query plan to the source for evaluation. This transfers all computation cost to the source, which is likely to be set up to handle such computations efficiently, e.g., with appropriate database, indices, and caches.

Unfortunately, this is not always the best course of action. As an example, consider a query that defines a Cartesian product of two base relations, such as the one shown below. For source relations with cardinality greater than 1, the result is larger than the individual relations.

```
query(X,Y) :- p(X) & q(Y)
```

At the other extreme, we can imagine asking the source for each relation individually. In this case, the broker does all of the work, and there is little burden on the source.

Unfortunately, this case also has problems. As an example, consider the following query, which requests the birthdate of all U.S. senators from the U.S. census database. There are just 100 senators, so the result here should be just 100 tuples. However, the size of the birthdate relation is in the hundreds of millions. Moving the birthdate database in this case would be foolhardy.

```
query(X,Y) :- senator(X) & birthdate(X,Y)
```

Statistics on the sizes of relations and estimates of join sizes can help us determine the most economical plan for retrieving the data. However, relation size statistics are not always available, and join size estimates, when available, are frequently unreliable.

The good news is that there is a reasonable rule for sourcing in the absence of such statistics and estimates. It does not produce optimal plans, but it guarantees that the cost of communication is no worse than a constant factor times the cost of communication in an optimal plan.

The basic idea of the rule is to break a retrievable query plan into maximal components in such a way that no component contains an unbounded join, i.e., a join of greater size than any of its inputs. Viewed from another perspective, the idea is to combine atomic queries into maximal conjunctions such that all conjunctions are intersections.

As an example of this rule, consider the following retrievable query plan.

```
query(X,Z) :- p(X,Y) & q(Y) & r(Y,Z)
```

The executable query plan shown below avoids joins that are not intersections.

```
query(X,Y) :- query_1(X,Y) & query_2(Y,Z)

query_1(X,Y) :- p(X,Y) & q(Y)
query_2(Y,Z) :- r(Y,Z)
```

Note that in some cases there are multiple executable query plans that satisfy the rule. For example, the following query plan works and also avoids joins that are not intersections.

```
query(X,Y) :- query_1(X,Y) & query_2(Y,Z)

query_1(X,Y) :- p(X,Y)
query_2(Y,Z) :- q(Y) & r(Y,Z)
```

Operationalizing a query plan in the presence of multiple data sources is more complicated than in the case of a single source. The main problem is replication of relations across multiple sources.

The presence of multiple copies of relations means that there are more equivalent executable query plans than there are when each relation has only one source.

The positive side of having these additional possibilities is that some of the extra plans may be more efficient than the plans obtainable without replication, e.g., by allowing more joins to be computed by sources without sending data from the broker.

The downside of having more executable query plans is that the data broker must sift through these additional possibilities to find an ideal executable query plan. The increased planning cost can offset any gains in execution efficiency, in some cases leading to a diminution in performance.

Computing an executable plan in the face of multiple sources is similar to that for computing executable plans with a single source. Before adding a conjunct to a query, the algorithm first checks that the conjunct can be evaluated by the source being considered for the other conjuncts in that group.

## 4.4 EXECUTION PLANNING

An executable query plan is more detailed than a retrievable query plan in that it specifies which data is to be gathered from which sources in order to answer a given overall query. However, it leaves open some questions about how this information is to be gathered from the sources. In this section, we look at some of the alternatives.

The simplest approach to plan execution is called *independent evaluation*. In this approach, the data broker sends the queries from the executable query plan to the relevant sources and then evaluates the local query plan using the answers that result from these queries.

As an example, consider a data integration setting in which there are two data sources (Alpha and Bravo) and four source relations (p, q, r, and s). Alpha stores p and q, and Bravo stores r and s. Assume that we are given the following executable query plan.

```
query(X,Z) :- alpha_1(X,Y) & bravo_1(Y,Z)

alpha_1(X,Z) :- p(X) & q(X,Y)
bravo_1(Y,Z) :- r(Y,Z) & s(Z)
```

Assume that the data shown below is stored at the Alpha site.

| p |
|---|
| a |
| b |
| c |
|   |

| q | |
|---|---|
| a | b |
| b | c |
| c | d |
| d | e |

And assume the following data is stored at the Bravo site.

| r | |
|---|---|
| a | e |
| b | f |
| c | g |
| d | h |

| s |
|---|
| e |
| f |
| g |
|   |

In independent evaluation, the first step would be for the broker to send Alpha the query shown below.

```
alpha_1(X,Y) :- p(X) & q(X,Y)
```

This would result in the following answer set.

| alpha_1 | |
| --- | --- |
| a | b |
| b | c |
| c | d |

The second step would be to send Bravo the following query.

```
bravo_1(Y,Z) :- r(Y,Z) & s(Z)
```

The result of this query is shown below.

| bravo_1 | |
| --- | --- |
| a | e |
| b | f |
| c | g |

The broker would then combine these tables in accordance with the local query plan to produce the following overall answer.

| query | |
| --- | --- |
| a | f |
| b | g |

One nice feature of independent evaluation is that queries can be evaluated in parallel. Thus, there is no need for everything to wait while a slow or temporarily busy site produces its answers.

The downside of independent evaluation is that all queries are evaluated in isolation; and, therefore, the method does not take advantage of inter-query relationships that might decrease the sizes of answer sets.

Imagine, for example, a query to find the ages of all U.S. senators, where the senators are mentioned in a congressional database and their ages are stored in a census database. Independent evaluation in this case requires that we retrieve all 100 senators from the congressional database. Okay, that is not so bad. However, it would also require that we retrieve the ages of everyone in the census database, producing an answer set with hundreds of millions of records when only 100 answers are needed.

The *semijoin method* attacks this problem by creating and sending queries that contain information about the values of variables obtained from the evaluation of other queries.

The simplest way to do this is for the broker to substitute the answers from any queries already evaluated into subsequent queries before evaluating those queries, thus creating a separate copy for each such answer. In our example, after evaluating the `alpha_1` query, the next step would be to send Bravo the following query.

```
bravo_1(b,Z)  :-  r(b,Z)  &  s(Z)
bravo_1(c,Z)  :-  r(c,Z)  &  s(Z)
bravo_1(d,Z)  :-  r(d,Z)  &  s(Z)
```

If the source accepts queries with embedded disjunctions, this could be expressed more compactly as shown below.

```
bravo_1(Y,Z)  :-  (Y=b | Y=c | Y=d) & r(Y,Z) & s(Z)
```

If the source accepts queries with set membership, the query could be expressed even more compactly as follows.

```
bravo_1(Y,Z)  :-  Y∈{b,c,d} & r(Y,Z) & s(Z)
```

Finally, in the unusual situation where the source allows the creation of temporary tables, the broker could create a table of known answers at the source and send an appropriate join query. In this case, the communication would look like the following. Here, we have named the temporary table `current` and populated it with the answers from `alpha_1`.

```
current(b)
current(c)
current(d)
bravo_1(Y,Z)  :-  current(Y) & r(Y,Z) & s(Z)
```

In any case, Bravo would return the answer shown below, which is a bit smaller than in the case of independent evaluation.

| bravo_1 | |
| --- | --- |
| b | f |
| c | g |

The good news about the semijoin method relative to independent evaluation is that it never increases the sizes of answer sets and often decreases those sizes. However, this decrease comes at the cost of larger queries that can offset this benefit. Let us look at the costs of the two approaches.

In independent evaluation, the total communication cost for evaluating a join of queries $A$ and $B$ is the sum of the costs of sending the queries and retrieving the answers. The size of each query is independent of the size of the database to which it is applied and hence can be ignored for all practical purposes. Hence, the communication cost is approximately the sum of the sizes of the answer sets $T_A$ and $T_B$.

$$T_A + T_B$$

The total communication cost for the semijoin method is shown below. The cost of the initial query can be ignored as before. The real cost is approximately the sum of the size of the initial answer set together with the size of the second query and the size of the second answer set. Here, we use $T_{A|B}$ to refer to the size of the answer set for $A$ restricted to the information referenced in $B$, and we let $T_{A|B\&B}$ refer to the resulting answer set for $B$.

$$T_A + T_{A|B} + T_{A|B\&B}$$

Clearly, the semijoin method is superior to independent evaluation when $(T_B - T_{A|B\&B})$ is greater than $T_{A|B}$. The good news about the semijoin method is that, even in the worst case, the cost is never more than a constant factor times the cost of independent evaluation, while there is a good chance that it can be much faster than independent evaluation.

One feature of both independent evaluation and the semijoin method is that the results must be joined with the results of other queries after the answers are obtained. This is annoying because these joins can sometimes be done more efficiently using indexes available in the sources.

An alternative that solves this problem is the *complete join method*. In this approach, *all* available information is passed along to the source, not just the information referenced in the query for that source. The source does the join and returns the full answer set. The broker then uses this result in constructing the query for the next source mentioned in the join.

In our example, the broker would include not just information about the variables in the source query but also information about the bindings of other variables needed in the final computation. In this case, it would send the following query (or one of its equivalents) to Bravo.

```
current(a,b)
current(b,c)
current(c,d)
query(X,Z) :- current(X,Y) & r(Y,Z) & s(Z)
```

Note that the variable X here corresponds to the variable X in the overall query, and the variable Y does not occur in the head as before.

Although this method can decrease computation cost, under certain circumstances, it can also increase communication and thereby offset any gains. The queries can be larger since more information must be passed along. Also, the resulting joins can end up being bigger than the answer sets for either of the other methods.

In the absence of key constraints and information about indexes, there is no way for a broker to know which of these approaches is best. A safe way to proceed is to use the semijoin method unless there is sufficient information to ensure the complete join method is superior.

·  ·  ·  ·

CHAPTER 5

# Master Schema Management

## 5.1 INTRODUCTION

The key to model-centric data integration is the availability of a good master schema. The schema must be rich enough to express the information in the data sources being integrated, but it should not be so elaborate as to make the data integration process needlessly inefficient.

The richness criterion can be made precise. Suppose we are given a schema $S$ with constraints $C_S$ and suppose we are given a master schema $M$ with constraints $C_M$. We say that $M$ is *adequate* for (defining) $S$ if and only if it is possible to write a definition for every relation $r$ in $S$ in terms of the relations in $M$ such that (1) for every instance of $M$ that satisfies $C_M$, the instance of $S$ given by these definitions satisfies the constraints in $C_S$ and (2) for every instance $D$ of $S$ that satisfies $C_S$, there is an instance of $M$ that satisfies $C_M$ and that yields $D$ under these definitions.

The efficiency criterion is more difficult to characterize, as it depends on the integration algorithm being used. In what follows, we discuss efficiency in the context of the algorithms described in the preceding chapters.

As an illustration of these ideas, consider two simple data schemas. The first consists of a single ternary relation q, and the second consists of a single ternary relation r. The first two fields of q and r are the same, while the third field in q differs from the third field of r.

We cannot use either one of these schemas as a master schema in this case, since it is not possible to define either schema in terms of the other.

Of course, we can use the union of the two schemas as our master schema so long as we add the constraints capturing the relationship between the two relations. In this case, the following constraints do the trick.

```
illegal :- q(U,V,W) & ~r(U,V,Z)
illegal :- ~q(U,V,W) & r(U,V,Z)
```

However, there is a simpler alternative. Consider a master schema with a single 4-ary relation p. The first two fields of p are the same as the first two fields in q and r, while the third field of p corresponds to the third field of q and the fourth field of p corresponds to the third field of r. The

*An interactive version of this book is available that includes implementations of algorithms used in demonstrations and exercises. See page xi for more information.*

good thing about this schema is that it is adequate for defining both q and r. The definitions are shown below.

```
q(U,V,W)  :- p(U,V,W,Z)
r(U,V,Z)  :- p(U,V,W,Z)
```

In the remaining sections of this chapter, we look at some ways in which master schemas can be written to improve the efficiency of data integration. Some ways are better in some cases and not so good in others. At this point in time, we do not know how to characterize a best schema for all data integration problems.

## 5.2 REIFICATION

Reification is the process of adding new objects to a schema and new relations on those objects to represent information previously expressed entirely with relations.

Consider a schema with three unary relations red, green, and blue. Each of these relations holds of an object if and only if the object has the corresponding color. Using this formulation, we might have a database like the one shown below.

```
red(a)
green(b)
blue(c)
```

Another way to represent color information is to reify colors as objects and to replace the various unary relations with a single binary relation color that associates objects with their colors. In this formulation, the data above would be rewritten as shown below.

```
color(a,red)
color(b,green)
color(c,blue)
```

Note that it is possible to take this process one step farther, reifying the concept of color as well and capturing data using a ternary relation property that relates a property (e.g., color), an object, and a value (e.g., color value). This formulation allows us to use a single relation to capture multiple properties (e.g., color, size, material, etc.).

```
property(color,a,red)
property(color,b,green)
property(color,c,blue)
```

One of the benefits of reification is that it sometimes simplifies the encoding of transformation rules and decreases the number of conjuncts in our query plans. As an example, consider the task of defining a relation r consisting of all pairs of objects satisfying a relation p such that the objects in each pair have different colors.

Using an unreified schema, such as the first shown above, and assuming there are just three colors, we would write our definition as shown below.

```
r(X,Y) :- p(X,Y) & red(X) & green(Y)
r(X,Y) :- p(X,Y) & red(X) & blue(Y)
r(X,Y) :- p(X,Y) & green(X) & red(Y)
r(X,Y) :- p(X,Y) & green(X) & blue(Y)
r(X,Y) :- p(X,Y) & blue(X) & red(Y)
r(X,Y) :- p(X,Y) & blue(X) & green(Y)
```

By contrast, using a reified schema, such as the second shown above, we could define our relation with a single rule.

```
r(X,Y) :- p(X,Y) & color(X,U) & color(Y,V) & U !=V
```

The benefit of reification becomes even more striking if we consider multiple properties. Suppose we wanted to define a relation consisting of all pairs of objects satisfying a binary relation p such that the objects in the pair have different colors and different sizes (drawn from the set small, medium, and large). In this case, using an unreified schema, we would have to consider all combinations of six color possibilities and six size possibilities, making 36 rules in all. By contrast, in a reified version, we would still need just one rule.

## 5.3    AUXILIARY TABLES

In many integration settings, it is valuable to provide auxiliary tables to express information as relational data rather than as rules. One benefit of this is increased execution performance, since using data is often faster than using rules. More significantly, this approach often cuts down on the number of query plans that must be considered at planning time.

As an example of using auxiliary tables, consider the problem of defining a table q of all objects in relation p together with their colors and sizes written in French, where the colors and sizes are defined via base tables written in English. The definition of q might look something like the following.

```
q(X,Y,Z) :- p(X) & couleur(X,Y) & taille(X,Z)
```

The definitions of `couleur` and `taille` are given below in terms of their English counterparts `color` and `size`.

```
couleur(X,rouge)  :- color(X,red)
couleur(X,noir)   :- color(X,black)
couleur(X,bleu)   :- color(X,blue)

taille(X,petit)   :- size(X,small)
taille(X,medium)  :- size(X,medium)
taille(X,grand)   :- size(X,large)
```

Doing the query planning in this case leads to the following query plan. There are nine conjuncts in the query plan.

```
q(X,rouge,petit)  :- p(X) & color(X,red) & size(X,small)
q(X,rouge,medium) :- p(X) & color(X,red) & size(X,medium)
q(X,rouge,grand)  :- p(X) & color(X,red) & size(X,large)
q(X,verd,petit)   :- p(X) & color(X,green) & size(X,small)
q(X,verd,medium)  :- p(X) & color(X,green) & size(X,medium)
q(X,verd,grand)   :- p(X) & color(X,green) & size(X,large)
q(X,bleu,petit)   :- p(X) & color(X,blue) & size(X,small)
q(X,bleu,medium)  :- p(X) & color(X,blue) & size(X,medium)
q(X,bleu,grand)   :- p(X) & color(X,blue) & size(X,large)
```

Now, consider an alternative formulation in which we invent tables of translations between words in French and English. We could invent similar tables for other languages such as Spanish.

```
e2f(red,rouge)
e2f(green,verd)
e2f(blue,bleu)

e2f(small,petit)
e2f(medium,medium)
e2f(large,grand)
```

Using these tables, we could define concepts like couleur and taille as follows.

```
couleur(X,Z)  :- color(X,Y) & e2f(Y,Z)

taille(X,Z)  :- size(X,Y) & e2f(Y,Z)
```

Using these auxiliary tables, query planning would produce the following query plan.

```
q(X,Y,Z) :- p(X) & color(X,U) & e2f(U,Y) & size(X,V) & e2f(V,Z)
```

Note that this plan is much smaller than before. Although execution requires a join between the color and size relations and the concordance relations, the overall cost is vastly reduced.

## 5.4   CONSTRAINT FOLDING

One criterion in evaluating a master schema is the degree of dependence among the base relations in the schema. The extreme of independence occurs when we have schemas in which all instances of all relations are possible.

The methods presented in Chapter 3 assume there are no constraints on the base relations in the master schema. In cases where this assumption does not hold, the answers are still correct but may be incomplete. In the presence of constraints, it is possible to obtain additional query foldings, some of which would not be correct in the absence of those constraints.

One way of dealing with this situation is *constraint folding*—we reformulate the master schema to take constraints into account and thereby make it possible for our constraint-independent query-folding methods to produce maximally contained plans.

Let us start with an extremely, almost trivial example to illustrate how constraints can destroy completeness of query folding and to see how we can reestablish completeness by building those constraints into our master schema.

Our master schema contains a ternary relation p with a constraint that the first and second arguments are identical.

```
illegal :- p(X,Y,Z) & X != Y
```

Let us assume there is a source relation v, which gives the projection of p onto its first and third arguments.

```
v(X,Z) :- p(X,Y,Z)
```

Now, consider a query asking for the projection of p onto its second and third arguments.

```
query(Y,Z) :- p(X,Y,Z)
```

Inverting our view in this case but without considering the constraints, we get the following.

```
p(X,f(X,Z),Z) :- v(X,Z)
```

Expanding our query with this inverse rule leads to the following.

$$\text{query}(f(Y,Z),Z) \; :- \; v(Y,Z)$$

While this plan is formally correct, all of its answers contain functional terms, and hence it is of no use.

The good news is that, given the constraint, we can convert our base relation p into a view by defining it in terms of an even more fundamental relation pb by adding in the constraint that the first two arguments are the same.

$$p(X,Y,Z) \; :- \; pb(X,Y,Z) \; \& \; X=Y$$

More succinctly, it could be written in the following form.

$$p(X,X,Z) \; :- \; pb(X,Z)$$

View inversion with this view definition leads to the following inverse rule.

$$pb(X,Z) \; :- \; v(X,Z)$$

Expanding our query with this new inverse rule leads to the following query plan.

$$\text{query}(Y,Z) \; :- \; v(Y,Z)$$

Our second example of constraint folding involves an inclusion dependency. In this case, we have two unary base relations, p and q with the constraint that p is contained in q.

$$\text{illegal} \; :- \; p(X) \; \& \; \sim q(X)$$

Let us assume that we have a single source relation v, which is identical to p.

$$v(X) \; :- \; p(X)$$

Now, let us assume we wish to query relation q.

$$\text{query}(X) \; :- \; q(X)$$

In the absence of the constraint, there is nothing we can do, since the query mentions q and there is no source relation that mentions q. However, taking the constraint into account, we realize that we can safely use v to answer the query.

We can get this result by folding in the constraint as shown below, where q is defined as the union of qb and p (due to the folding constraint).

```
q(X)  :- qb(X)
q(X)  :- p(X)
```

The inverse of our view definition is shown below.

```
p(X)  :- v(X)
```

Using these definitions, query folding leads to the following plan, which is maximally contained in the original query given the constraint.

```
query(X)  :- v(X)
```

It is worth noting that the constraint-folding technique described here can be used during the query-folding process. However, folding in constraints in advance has the merit of saving a small amount of computation at runtime. It also allows us to separate the constraint-folding and query-folding code, thus simplifying our data integration code.

·   ·   ·   ·

# Appendix

## A.1 INTRODUCTION

The code presented here is a JavaScript implementation of the algorithms described in the text. It is a reference implementation, written for readability/understandability, not efficiency.

While the code can be used as written, there are numerous ways it can be improved. Representing entities, relations, and variables as unique data structures can decrease the cost of comparisons. Avoiding the copying of sequences can also help.

The code here has been used in a wide variety of demonstrations and appears to work without flaws. However, only limited quality assurance methodology has been applied; thus, there may be cases where the code does not behave as expected. Reader, beware.

## A.2 SENTENTIAL REPRESENTATION

Entities, relations, and variables are represented by strings of characters following the conventions in the text. Propositions/atomic sentences are represented as sequences of strings with the relation as the first element and the arguments as the succeeding elements. Complex expressions in Datalog are represented as sequences of component expressions, with the operator as the first element and the components as the succeeding elements. The following subroutines are used to test the types of expressions and to construct expressions of various types. There are also subroutines for computing various operations on expressions and sequences of expressions.

```
function symbolp (x)
  {return typeof x == 'string'}

function varp (x)
  {return typeof x == 'string' &&
         x.length != 0 &&
         x[0] != x[0].toLowerCase()}
```

*An interactive version of this book is available that includes implementations of algorithms used in demonstrations and exercises. See page xi for more information.*

```
function constantp (x)
 {return typeof x == 'string' &&
         x.length != 0 &&
         x[0] == x[0].toLowerCase()}

var counter = 0

function newvar ()
 {counter++;  return 'V' + counter}

function newsym ()
 {counter++;  return 'c' + counter}

function seq ()
 {var exp=new Array(arguments.length);
  for (var i=0; i<arguments.length; i++)
{exp[i]=arguments[i]};
  return exp}

function head (p)
 {return p[0]}

function tail (l)
 {return l.slice(1,l.length)}

function makeequality (x,y)
 {return seq('same',x,y)}

function makeinequality (x,y)
 {return seq('distinct',x,y)}

function makenegation (p)
 {return seq('not',p)}

function makeconjunction (p,q)
 {if (p[0] == 'and') {return p.concat(seq(q))};
  return seq('and',p,q)}
```

```
function makedisjunction (p,q)
 {if (p[0] == 'or') {return p.concat(seq(q))};
  return seq('or',p,q)}

function makereduction (head,body)
 {return seq('reduction',head,body)}

function makeimplication (head,body)
 {return seq('implication',head,body)}

function makeequivalence (head,body)
 {return seq('equivalence',head,body)}

function makerule (head,body)
 {if (body.length == 0) {return head};
  if (body[0] == 'and') {return
seq('rule',head).concat(tail(body))};
   return seq('rule',head,body)}

function makeuniversal (variable,scope)
 {return seq('forall',variable,scope)}

function makeexistential (variable,scope)
 {return seq('exists',variable,scope)}

function makeconditional (p,x,y)
 {return seq('if',p,x,y)}

function makeclause (p,q)
 {return seq('clause',p,q)}

function makedefinition (head,body)
 {if (!symbolp(body) & body[0]=='and')
     {return seq('definition',head).concat(tail(body))}
   else {return seq('definition',head,body)}}
```

```
function makestep (sentence,justification,p1,p2)
 {var exp = new Array(3);
  exp[0] = 'step';
  exp[1] = sentence;
  exp[2] = justification;
  if (p1) {exp[3] = p1};
  if (p2) {exp[4] = p2};
  return exp}

function makeproof ()
 {var exp = new Array(1);
  exp[0] = 'proof';
  return exp}

function maksatom (r,s)
 {return seq(r).concat(s)}

function maksand (s)
 {if (s.length == 0) {return 'true'};
  if (s.length == 1) {return s[0]};
  return seq('and').concat(s)}

function maksor (s)
 {if (s.length == 0) {return 'false'};
  if (s.length == 1) {return s[0]};
  return seq('or').concat(s)}

function negate (p)
 {if (symbolp(p)) {return makenegation(p)};
  if (p[0] == 'not') {return p[1]};
  return makenegation(p)}

function adjoin (x,s)
 {if (!findq(x,s)) {s.push(x)};
  return s}
```

```
function concatenate (l1,l2)
 {return l1.concat(l2)}

function findq (x,s)
 {for (var i=0; i<s.length; i++) {if (x == s[i]) {return true}};
  return false}

function find (x,s)
 {for (var i=0; i<s.length; i++) {if (equalp(x,s[i]))
{return true}};
  return false}

function subset (s1,s2)
 {for (var i=0; i<s1.length; i++)
     {if (!find(s1[i],s2)) {return false}};
  return true}

function difference (l1, l2)
 {var answer = seq();
  for (var i=0; i<l1.length; i++)
     {if (!find(l1[i],l2)) {answer[answer.length] = l1[i]}};
  return answer}

function subst (x,y,z)
 {if (z == y) {return x};
  if (symbolp(z)) {return z};
  var exp = new Array(z.length);
  for (var i=0; i<z.length; i++)
     {exp[i] = subst(x,y,z[i])};
  return exp}

function substitute (p,q,r)
 {if (symbolp(r)) {if (r == p) {return q} else {return r}};
  var exp = seq();
  for (var i=0; i<r.length; i++)
```

```
        {exp[exp.length] = substitute(p,q,r[i])};
   if (equalp(exp,p)) {return q} else {return exp}}


function substitutions (p,q,r)
 {if (symbolp(r))
     {if (r == p) {return seq(r,q)} else {return seq(r)}};
   return substitutionsexp(p,q,r,0)}


function substitutionsexp (p,q,r,n)
 {if (n == r.length) {return seq(seq())};
  var firsts = substitutions(p,q,r[n]);
  var rests = substitutionsexp(p,q,r,n+1);
  var results = seq();  for (var i=0; i<firsts.length; i++)
     {for (var j=0; j<rests.length; j++)
          {exp = seq(firsts[i]).concat(rests[j]);
           results[results.length] = exp;
           if (equalp(exp,p)) {results[results.length] = q}}}
   return results}


function vars (x)
 {return varsexp(x,seq())}


function varsexp (x,vs)
 {if (varp(x)) {return adjoin(x,vs)};
  if (symbolp(x)) {return vs};
  for (var i=0; i<x.length; i++) {vs = varsexp(x[i],vs)};
  return vs}


function constants (x)
 {return constantsexp(x,seq())}


function constantsexp (x,vs)
 {if (varp(x)) {return vs};
  if (symbolp(x)) {return adjoin(x,vs)};
  for (var i=1; i<x.length; i++) {vs = constantsexp(x[i],vs)};
  return vs}
```

```
function equalp (p,q)
 {if (symbolp(p)) {if (symbolp(q)) {return p==q} else {return
      false}};
  if (symbolp(q)) {return false};
  if (p.length != q.length) {return false};
  for (var i=0; i<p.length; i++) {if (!equalp(p[i],q[i])) {return
      false}};
  return true}
```

## A.3 LINKED LISTS

Some of our algorithms utilize linked lists rather than sequences. The following subroutines define constructors and accessors on linked lists and various other operations.

```
var nil = 'nil'

function nullp (l)
 {return l == 'nil'}

function cons (x,l)
 {var cell = new Array(2);
  cell[0] = x;
  cell[1] = l;
  return cell}

function car (l)
 {return l[0]}

function cdr (l)
 {return l[1]}

function list ()
 {var exp=nil;
  for (var i=arguments.length; i>0; i--)
      {exp=cons(arguments[i-1],exp)};
  return exp}
```

```
function len (l)
  {var n = 0;
   for (var m=l; m!=nil; m = cdr(m)) {n = n+1};
   return n}

function memberp (x,l)
  {if (nullp(l)) {return false};
   if (equalp(car(l),x)) {return true};
   if (memberp(x,cdr(l))) {return true};
   return false}

function append (l1,l2)
  {if (nullp(l1)) {return l2}
      else {return cons(car(l1),append(cdr(l1),l2))}}

function nreverse (l)
  {if (nullp(l)) {return nil}
   else {return nreversexp(l,nil)}}

function nreversexp (l,ptr)
  {if (cdr(l) == nil) {l[1] = ptr; return l};
   var rev = nreversexp(cdr(l),l);
   l[1] = ptr;
   return rev}

function acons (x,y,al)
  {return cons(cons(x,y),al)}

function assoc (x,al)
  {if (nullp(al)) {return false};
   if (x == car(car(al))) {return car(al)};
   return assoc(x,cdr(al))}
```

## A.4   UNIFICATION

The following subroutines check for matching of expressions and unification of expressions.
There are also subroutines for applying substitutions to expressions and renaming variables within
expressions.

```
function matcher (x,y)
 {return match(x,y,nil)}

function match (x,y,bl)
 {if (x == y) {return bl};
  if (varp(x)) {return matchvar(x,y,bl)};
  if (symbolp(x)) {return false};
  return matchexp(x,y,bl)}

function matchvar (x,y,bl)
 {var dum = assoc(x,bl);
  if (dum != false) {return match(cdr(dum),y,bl)};
  if (x == matchval(y,bl)) {return bl};
  return acons(x,y,bl)}

function matchval (y,bl)
 {if (varp(y))
     {var dum = assoc(y,bl);
      if (dum != false) {return matchval(cdr(dum),bl)};
      return y};
   return y}

function matchexp(x,y,bl)
 {if (symbolp(y)) {return false};
  var m = x.length;
  var n = y.length;
  if (m != n) {return false};
  for (var i=0; i<m; i++)
     {bl = match(x[i],y[i],bl);
      if (bl == false) {return false}};
   return bl}

function unifier (x,y)
 {return unify(x,y,nil)}

function unify (x,y,bl)
 {if (x == y) {return bl};
```

```
      if (varp(x)) {return unifyvar(x,y,bl)};
      if (symbolp(x)) {return unifyatom(x,y,bl)};
      return unifyexp(x,y,bl)}

 function unifyvar (x,y,bl)
  {var dum = assoc(x,bl);
   if (dum != false) {return unify(cdr(dum),y,bl)};
   if (x == unifyval(y,bl)) {return bl};
   return acons(x,y,bl)}

 function unifyval (y,bl)
  {if (varp(y))
      {var dum = assoc(y,bl);
       if (dum != false) {return unifyval(cdr(dum),bl)};
       return y};
   return y}

 function unifyatom (x,y,bl)
  {if (varp(y)) {return unifyvar(y,x,bl)} else return false}

 function unifyexp(x,y,bl)
  {if (varp(y)) {return unifyvar(y,x,bl)}
   if (symbolp(y)) {return false};
   if (x.length != y.length) {return false};
   for (var i=0; i<x.length; i++)
      {bl = unify(x[i],y[i],bl);
       if (bl == false) {return false}};
   return bl}

 function plug (x,bl)
  {if (varp(x)) {return plugvar(x,bl)};
   if (symbolp(x)) {return x};
   return plugexp(x,bl)}

 function plugvar (x,bl)
  {var dum = assoc(x,bl);
   if (dum == false) {return x};
   return plug(cdr(dum),bl)}
```

```
function plugexp (x,bl)
 {var exp = new Array(x.length);
  for (var i=0; i<x.length; i++)
      {exp[i] = plug(x[i],bl)};
  return exp}


var alist;


function standardize (x)
 {alist = nil;
  return standardizeit(x)}


function standardizeit (x)
 {if (varp(x)) {return standardizevar(x)};
  if (symbolp(x)) {return x};
  return standardizeexp(x)}


function standardizevar (x)
 {var dum = assoc(x,alist);
  if (dum != false) {return cdr(dum)};
  var rep = newvar();
  alist = acons(x,rep,alist);
  return rep}


function standardizeexp (x)
 {var exp = new Array(x.length);
  for (var i=0; i<x.length; i++)
      {exp[i] = standardizeit(x[i])};
  return exp}
```

## A.5   STORAGE

In our implementation here, we represent rules in one "theory" and data in a second theory. In each case, theory is represented as a sequence of sentences. Indexing is supported using Javascript's associative arrays.

```
var indexing = true
```

```
function definetheory (source,data)
 {emptytheory(source);
  definemore(source,data);
  return true}

function definemore (theory,data)
 {for (var i=0; i<data.length; i++) {insert(data[i],theory)};
  return true}

function emptytheory (theory)
 {theory.splice(0,theory.length);
  for (var x in theory) {delete theory[x]};
  return true}

function drop (p,theory)
 {data = indexps(p,theory);
  for (var i=0; i<data.length; i++)
      {if (equalp(data[i],p))
           {uninsert(data[i],theory); return data[i]}};
  return false}

function eliminate (object,theory)
 {var data = indexees(object,theory).concat();
  for (var i=0; i<data.length; i++)
      {if (data[i][1] == object) {uninsert(data[i],theory)}};
  return object}

function insert(p,theory)
 {addcontent(p,theory);
  if (indexing) {index(p,p,theory)};
  return p}

function addcontent (p,theory)
 {theory.push(p);
  return p}

function index (x,p,theory)
 {if (varp(x)) {return p};
```

```
   if (symbolp(x)) {return indexsymbol(x,p,theory)};
   for (var i=0; i<x.length; i++) {index(x[i],p,theory)};
   return p}

function indexsymbol (x,p,theory)
 {if (x == 'map') {return p};
  if (!isNaN(Number(x))) {return p};
  var data = theory[x];
  if (data) {data.push(p)} else {theory[x] = seq(p)};
  return p}

function uninsert(p,theory)
 {if (indexing) {unindex(p,p,theory)};
  return remcontent(p,theory)}

function remcontent (p,theory)
 {for (var i=0; i<theory.length; i++)
     {if (theory[i]==p) {theory.splice(i,1); return p}};
  return false}

function unindex (x,p,theory)
 {if (varp(x)) {return p};
  if (symbolp(x)) {return unindexsymbol(x,p,theory)};
  for (var i=0; i<p.length; i++) {unindex(x[i],p,theory)};
  return p}

function unindexsymbol (x,p,theory)
 {if (theory[x]) {return remcontent(p,theory[x])}}

function indexps (p,theory)
 {if (indexing) {return fullindexps(p,theory)};
  return theory}

function fullindexps (p,theory)
 {if (varp(p)) {return theory};
  if (symbolp(p)) {return indexees(p,theory)};
  for (var i=1; i<p.length; i++)
```

```
          {if (symbolp(p[i]) && !varp(p[i]))
             {return indexees(p[i],theory)}};
    return indexees(p[0],theory)}

function indexees (x,theory)
 {if (indexing && x != 'map' && isNaN(Number(x)))
     {var data = theory[x];
      if (data) {return data} else {return seq()}};
   return theory}

function envindexps (p,al,theory)
 {if (indexing) {return dataindexps(p,al,theory)};
   return theory}

function dataindexps (p,al,theory)
 {if (varp(p)) {return theory};
  if (symbolp(p)) {return indexees(p,theory)};
  for (var i=1; i<p.length; i++)
      {var dum = unival(p[i],al);
       if (symbolp(dum) && !varp(dum))
          {return indexees(dum,theory)}};
   return indexees(p[0],theory)}

function unival (x,al)
 {if (!varp(x)) {return x};
  var dum = assoc(x,al);
  if (dum) {return unival(cdr(dum),al)};
  return x}

function uniquify (ins)
 {var outs = seq();
  for (var i=0; i<ins.length; i++) {outs = adjoinit(ins[i],outs)};
  return outs}

function adjoinit (x,s)
 {if (find(x,s)) {return s} else {return concatenate(s,seq(x))}}
```

```
function nconc (l1,l2)
  {for (var i=0; i<l2.length; i++) {l1.push(l2[i])};
   return l1}
```

## A.6  LOCAL EVALUATION

The subroutines in this section implement evaluation of open Datalog programs on extensional databases. The program is assumed to be stored in the global variable `rulebase` and the data is assumed to be stored in the global variable `database`. The method used here is top–down evaluation. This is appropriate for small databases and cases where the user can be assured that there are no unbounded recursions. For other cases, bottom–up evaluation is superior.

Note that there are two versions of the subroutines here: one for computing just one answer to the query, the other for computing all answers. While users seldom ask for the former, it is useful in computing negation as failure. If there is one answer to a subquery, then any negation of that subquery must be false and so computation on that thread of evaluation can be terminated.

```
var thing;
var answer;
var answers;

function findp (query,facts,rules)
  {return findx('true',query,facts,rules)}

function findx (result,query,facts,rules)
  {thing = result;
   answer = false;
   if (proone(query,seq(),nil,facts,rules)) {return answer};
   return false}

function finds (result,query,facts,rules)
  {thing = result;
   answers = seq();
   proall(query,seq(),nil,facts,rules);
   return uniquify(answers)}

function proone (p,pl,al,facts,rules)
  {if (symbolp(p)) {return prooneatom(p,pl,al,facts,rules)}
```

```
    if (p[0] == 'same')
       {return proonesame(p,pl,al,facts,rules)}
    if (p[0] == 'distinct')
       {return proonedistinct(p,pl,al,facts,rules)}
    if (p[0] == 'not')
       {return proonenot(p,pl,al,facts,rules)}
    if (p[0] == 'and')
       {return prooneand(p,pl,al,facts,rules)}
    if (p[0] == 'or')
       {return prooneor(p,pl,al,facts,rules)}
    if (proonebackground(p,pl,al,facts,rules))
       {return true};
    return prooners(p,pl,al,facts,rules)}

function prooneatom (p,pl,al,facts,rules)
 {if (p == 'true') {return prooneexit(pl,al,facts,rules)};
  if (p == 'false') {return false};
   return prooners(p,pl,al,facts,rules)}

function proonenot (p,pl,al,facts,rules)
 {if (proone(p[1],seq(),al,facts,rules) == false)
       {return prooneexit(pl,al,facts,rules)}
   return false}

function prooneand (p,pl,al,facts,rules)
 {return prooneexit(concatenate(tail(p),pl),al,facts,rules)}

function prooneor (p,pl,al,facts,rules)
 {var bl;
   for (var i=0; i<p.length; i++)
       {if (bl == proone(p[i],pl,al,facts,rules)) {return true}}
    return false}

function proonesame (p,pl,al,facts,rules)
 {al = unify(p[1],p[2],al);
   if (al != false) {return prooneexit(pl,al,facts,rules)};
   return false}
```

```
function proonedistinct (p,pl,al,facts,rules)
 {if (unify(p[1],p[2],al) == false)
     {return prooneexit(pl,al,facts,rules)};
  return false}

function proonebackground (p,pl,al,facts,rules)
 {var bl;
  var data = envindexps(p,al,facts);
  for (var i=0; i<data.length; i++)
     {bl = unify(data[i],p,al);
      if (bl != false && prooneexit(pl,bl,facts,rules))
         {return true}};
  return false}

function prooners (p,pl,al,facts,rules)
 {var copy;
  for (var i=0; i<rules.length; i++)
     {copy = standardize(rules[i]);
      if (copy[0] == 'rule')
         {var bl = unify(copy[1],p,al);
          var cont = concatenate(copy.slice(3),pl)
          if (bl != false &&
              proone(copy[2],cont,bl,facts,rules))
             {return true}}
         else {var bl = unify(copy,p,al);
               if (bl != false &&
                   prooneexit(pl,bl,facts,rules))
                  {return true}}};
   return false}

function prooneexit (pl,al,facts,rules)
 {if (pl.length != 0)
     {return proone(pl[0],tail(pl),al,facts,rules)};
  answer = plug(thing,al);
  return true}
```

```
function proall (p,pl,al,facts,rules)
  if (symbolp(p)) {return proallatom(p,pl,al,facts,rules)};
  if (p[0] == 'same')
     {return proallsame(p,pl,al,facts,rules)}
  if (p[0] == 'distinct')
     {return proalldistinct(p,pl,al,facts,rules)}
  if (p[0] == 'matches')
     {return proallmatches(p,pl,al,facts,rules)}
  if (p[0] == 'not')
     {return proallnot(p,pl,al,facts,rules)}
  if (p[0] == 'and')
     {return proalland(p,pl,al,facts,rules)}
  if (p[0] == 'or')
     {return proallor(p,pl,al,facts,rules)}
  proallbackground(p,pl,al,facts,rules);
  return proallrs(p,pl,al,facts,rules)}

function proallatom (p,pl,al,facts,rules)
 {if (p == 'true') {return proallexit(pl,al,facts,rules)};
  if (p == 'false') {return false};
  return proallrs(p,pl,al,facts,rules)}

function proallsame (p,pl,al,facts,rules)
 {al = unify(p[1],p[2],al);
  if (al != false) {proallexit(pl,al,facts,rules)}}

function proalldistinct (p,pl,al,facts,rules)
 {if (unify(p[1],p[2],al) == false)
     {proallexit(pl,al,facts,rules)}}

function proallmatches (p,pl,al,facts,rules)
 {if (symbolp(p[1]))
     {var matches = p[1].match(p[2]);
      for (var i=0; i<matches.length; i++)
          {var bl = unify(p[3],matches[i],al);
           if (bl != false)
              {proallexit(pl,bl,facts,rules)}}}
  return false}
```

```
function proallnot (p,pl,al,facts,rules)
 {if (proone(p[1],seq(),al,facts,rules) == false)
     {proallexit(pl,al,facts,rules)}}


function proalland (p,pl,al,facts,rules)
 {proallexit(concatenate(tail(p),pl),al,facts,rules)}


function proallor (p,pl,al,facts,rules)
 {for (var i=0; i<p.length; i++)
     {proall(p[i],pl,al,facts,rules)}}


function proallbackground (p,pl,al,facts,rules)
 {var bl;
  var data = envindexps(p,al,facts);
  for (var i=0; i<data.length; i++)
     {bl = match(p,data[i],al);
      if (bl != false) {proallexit(pl,bl,facts,rules)}}}


function proallrs (p,pl,al,facts,rules)
 {var copy;
  var bl;
  for (var i=0; i<rules.length; i++)
     {copy = standardize(rules[i]);
      if (copy[0] == 'rule')
          {bl = unify(copy[1],p,al);
           if (bl != false)
              {var cont = concatenate(copy.slice(3),pl);
               proall(copy[2],cont,bl,facts,rules)}}
      else {bl = unify(copy,p,al);
            if (bl != false) {proallexit(pl,bl,facts,rules)}}}}


function proallexit (pl,bl,facts,rules)
 {if (pl.length != 0)
     {return proall(pl[0],tail(pl),bl,facts,rules)};
  answers.push(plug(thing,bl))}
```

## A.7    QUERY FOLDING

The following subroutines compute the inverse for a single conjunctive view definition.

```
function inversesall (vs)
 {var rules = seq();
  for (var i=0; i<vs.length; i++)
      {rules = rules.concat(inverses(vs[i]))};
   return rules}


function inverses (v)
 {var rules = seq();
  var sigma = seq();
  var hs = vars(v[1]);
  var xs = difference(vars(v),hs);
  for (var i=0; i<xs.length; i++)
      {sigma[xs[i]] = maksatom(newsym(),hs)};
  for (var j=2; j<v.length; j++)
      {rules[rules.length] = makerule(rewrite(v[j],sigma),v[1])};
   return rules}


function rewrite (x,bl)
 {if (varp(x)) {return rewritevar(x,bl)};
  if (symbolp(x)) {return x};
  return rewriteexp(x,bl)}


function rewritevar (x,bl)
 {var dum = bl[x];
  if (dum == null) {return x};
  return rewrite(dum,bl)}


function rewriteexp (x,bl)
 {var exp = new Array(x.length);
  for (var i=0; i<x.length; i++)
      {exp[i] = rewrite(x[i],bl)};
        return exp}
```

The subroutine `flattenall` takes a rule set as argument and returns the result of applying this transformation.

```
function flattenatom (p)
 {var name = p[0];
  var exp = seq(name);
  for (var i=1; i<p.length; i++)
      {if (symbolp(p[i]) || varp(p[i]))
           {name = name + i;
            exp[exp.length] = p[i]}
        else {name = name + p[i][0];
               for (var j=0; j<p[i].length; j++)
                    {var n = i+j;
                     name = name + n;
                     exp[exp.length] = p[i][j]}}};
  exp[0] = name;
  return exp}


function flatten (r)
 {return seq('rule',flattenatom(r[1])).concat(r.slice(2))}


function flattenatom (p)
 {var name = p[0];
  var exp = seq(name);
  for (var i=1; i<p.length; i++)
      {if (symbolp(p[i]) || varp(p[i]))
           {name = name + i;
            exp[exp.length] = p[i]}
        else {name = name + p[i][0];
               for (var j=0; j<p[i].length; j++)
                    {var n = i+j;
                     name = name + n;
                     exp[exp.length] = p[i][j]}}};
  exp[0] = name;
  return exp}
```

## A.8    SOURCING

The `multisource` subroutine shown below is a greedy algorithm that adds in as many conjuncts as it can subject to the additional constraint that all conjuncts can be handled by a single source. If there are multiple sources that can handle all of the conjuncts in a group, it chooses arbitrarily.

```
function multisource (rule)
 {var toprule = seq(rule[0],rule[1]);
 var newrule;
 var newrules = seq(toprule);
 var done = seq();
 var sources;
 var temp;
 var n = 0;
 for (i=2; i<rule.length; i++)
     {var dum = seq('query_' + n++).concat(vars(rule[i]));
      newrule = seq(rule[0],dum);
      sources = specialists(head(rule[i]));
      if (sources.length == 0) {return false};
      if (!find(rule[i],done))
         {done[done.length] = rule[i];
          newrule[newrule.length] = rule[i];
          for (j=i+1; j<rule.length; j++)
              {if (!find(rule[j],done))
                  {temp=intersection(sources,specialists(head(rule[j])));
                   if (temp.length != 0 &&
                       (subset(vars(rule[j]),vars(rule[i])) ||
                        subset(vars(rule[i]),vars(rule[j]))))
                      {sources = temp;
                       done[done.length] = rule[j];
                       newrule[newrule.length] = rule[j]}}};
          newrule[1][0] = sources[0] + '_' + n++;
          toprule[toprule.length]=newrule[1];
          newrules[newrules.length] = newrule}};
 return newrules}
```

```
function intersection (s1, s2)
 {var result = seq();
  for (var i=0; i<s1.length; i++)
      {if (find(s1[i],s2))
          {result[result.length] = s1[i]}};
  return result}
```

Note that the code assumes subroutine `specialists` that takes a relation as argument and returns a sequence of sources that contain that relation.

# References

Serge Abiteboul, Oliver Duschka: Complexity of Answering Queries Using Materialized Views. *Proceedings of the Seventeenth ACM Symposium on Principles of Database Systems*, 1998. doi:10.1145/275487.275516

Ashok Chandra, Philip Merlin: Optimal Implementation of Conjunctive Queries in Relational Databases. *Proceedings of the Ninth ACM Symposium on the Theory of Computing*, pp. 77–90, 1977.

Surajit Chaudhuri, Ravi Krishnamurthy, Spyros Potamianos, Kyuseok Shim: Optimizing Queries with Materialized Views. *Proceedings of the 11th International Conference on Data Engineering*, pp. 190–200, 1995. doi:10.1109/ICDE.1995.380392

Rada Chirkova, Michael Genesereth: Linearly Bounded Reformulations of Unary Databases. *Symposium on Abstraction, Reformulation, and Approximation*, 2000. doi:10.1007/3-540-44914-0_9

Rada Chirkova, Michael Genesereth: Linearly Bounded Reformulations of Conjunctive Databases. *International Conference of Deductive and Object-Oriented Databases*, 2000. doi:10.1007/3-540-44957-4_66

Rada Chirkova, Michael Genesereth: Database Reformulation with Integrity Constraints. *The Logic and Computational Complexity Workshop*, in conjunction with the *Conference on Logic in Computer Science*, 2005.

Oliver Duschka, Michael Genesereth: Infomaster—An Information Integration Tool. *Proceedings of the International Workshop on Intelligent Information Integration, 21st German Annual Conference on Artificial Intelligence*, 1997.

Oliver Duschka, Michael Genesereth: Query Planning in Infomaster. *Proceedings of the ACM Symposium on Applied Computing*, 1997. doi:10.1145/331697.331719

Oliver Duschka, Alon Levy: Recursive Plans for Information Gathering. *Proceedings of the Fifteenth International Joint Conference on Artificial Intelligence*, 1997. doi:10.1145/263661.263674

Oliver Duschka, Michael Genesereth: Answering Recursive Queries Using Views. *Proceedings of the Sixteenth ACM Symposium on Principles of Database Systems*, pp. 109–116, 1997.

Oliver Duschka, Michael Genesereth: Query Planning with Disjunctive Sources. *Proceedings of the AAAI Workshop on AI and Information Integration*, 1998.

Oliver Duschka, Michael Genesereth, Alon Levy: Recursive Query Plans for Data Integration. *Journal of Logic Programming, Special Issue on Logic Based Heterogeneous Information Systems*, 1999.

Thomas Eiter, Georg Gottlob, Heikki Mannila: Adding Disjunction to Datalog. *Proceedings of the 13th ACM Symposium on Principles of Database Systems*, pp. 267–278, 1994. doi:10.1145/182 591.182639

Jarek Gryz: Query Folding with Inclusion Dependencies. *Proceedings of the 14th International Conference on Data Engineering*, pp. 126–133, 1999.

Tomas Imielinski, Witold Lipski Jr.: Incomplete Information in Relational Databases. *Journal of Association for Computing Machinery*, Volume 31, Number 4, pp. 761–791, 1984. doi:10.1145/1634.1886

Richard Karp: Reducibility Among Combinatorial Problems. *Complexity of Computer Computations*, pp. 85–104, 1972.

Thomas Kirk, Alon Levy, Yehoshua Sagiv, Divesh Srivastava: The Information Manifold. *Proceedings of the AAAI Spring Symposium on Information Gathering in Distributed Heterogeneous Environments*, 1995.

Chung Kwok, Daniel Weld: Planning to Gather Information. *Proceedings of the 13th National Conference on Artificial Intelligence*, 1996.

Alon Levy, Divesh Srivastava, Thomas Kirk: Data Model and Query Evaluation in Global Information Systems. *Journal of Intelligent Information Systems: Integrating Artificial Intelligence and Database Technologies*, Volume 5, Number 2, pp. 121–143, 1995. doi:10.1007/BF00962627

Alon Levy, Alberto Mendelzon, Yehoshua Sagiv, Divesh Srivastava: Answering Queries Using Views. *Proceedings of the ACM Symposium on Principles of Database Systems*, 1995. doi:10.1145 /212433.220198

Alon Levy, Anand Rajaraman, Joann Ordille: Querying Heterogeneous Information Sources Using Source Descriptions. *Proceedings of the 22nd International Conference on Very Large Databases*, pp. 251–262, 1996.

Alon Levy, Anand Rajaraman, Jeffrey Ullman. Answering Queries Using Limited External Processors. *Proceedings of the 15th ACM Symposium on Principles of Database Systems*, 1996. doi:10.1 145/237661.237716

Rachel Pottinger, Alon Halevy: MiniCon: A Scalable Algorithm for Answering Queries Using Views. *VLDB Journal*, Volume 10, Numbers 2–3, pp. 182–198, 2001.

Xiaolei Qian: Query Folding. *Proceedings of the 12th International Conference on Data Engineering*, pp. 48–55, 1996. doi:10.1109/ICDE.1996.492088

Anand Rajaram, Yehoshua Sagiv, Jeffrey Ullman: Answering Queries Using Templates with Binding Patterns. *Proceedings of the 14th ACM Symposium on Principles of Database Systems*, pp. 172–181, 1989.

Oded Shmueli: Decidability and Expressiveness Aspects of Logic Queries. *Proceedings of the ACM Symposium on Principles of Database Systems*, pp. 237–249, 1987. doi:10.1145/28659.28685

Devika Subramanian, Michael Genesereth: The Relevance of Irrelevance. *Proceedings of the International Joint Conference on Artificial Intelligence*, 1987.

Jeffrey Ullman: Information Integration Using Logical Views. *Proceedings of the Sixth International Conference on Database Theory*, pp. 19–40, 1997.

Jeffrey Ullman: *Principles of Database and Knowledgebase Systems*, Volume 2, Computer Science Press, Potomac, MD, 1989.

Moshe Vardi: The Complexity of Relational Query Languages. *Proceedings of the 14th Annual ACM Symposium on the Theory of Computing*, pp. 137–146, 1982. doi:10.1145/800070.802186

H. Z. Yang, Paul Larson: Query Transformation for PSJ Queries. *Proceedings of the 13th International Conference on Very Large Data Bases*, pp. 245–254, 1987.

# Index

# Author Biography

**Michael Genesereth** is an associate professor in the Computer Science Department at Stanford University. He received his Sc.B. in Physics from MIT and his Ph.D. in Applied Mathematics from Harvard University. Prof. Genesereth is most known for his work on computational logic and applications of that work in enterprise computing and electronic commerce. His current research interests include data integration and dissemination, enterprise management, and computational law. Prof. Genesereth was program chairman for the 1983 AAAI Conference and the 1997 International World Wide Web Conference. He is one of the founders of Teknowledge, the premier company commercializing Artificial Intelligence; he is a cofounder of CommerceNet, the premier organization for electronic commerce on the Internet; and he is a founder of Mergent Systems, an early vendor of technology for integrated catalogs on the World Wide Web. He is the current director of the Logic Group at Stanford and founder and research director of CodeX (The Stanford Center for Computers and Law).

Printed in the United States
by Baker & Taylor Publisher Services